ASPECTS OF CHANGE IN SENECA IROQUOIS LADLES A.D. 1600-1900

by
Betty Coit Prisch
Drawings by Gene Mackay

supported by the
Arthur C. Parker Fund for Iroquois Research

Research Records No. 15
1982

GENERAL EDITOR
Charles F. Hayes III

ASSOCIATE EDITOR
Ann Prichard

published by the
RESEARCH DIVISION

Rochester Museum and Science Center
657 East Avenue, Rochester, New York 14603

The articles in this volume do not represent any consensus of beliefs. Views expressed are those of the author and do not necessarily reflect the position of the publisher.

Copyright 1982

Rochester Museum & Science Center
657 East Avenue
Rochester, New York 14603
Richard C. Shultz, President

Library of Congress Catalog Card Number: 82-61498

Printed in U.S.A.

Printed by Monroe Reprographics, Inc., Rochester, NY

CONTENTS

Preface	*Charles F. Hayes III*	v
Acknowledgements	*Betty Coit Prisch*	vii
General Introduction		1
Chapter I Seventeenth Century Archaeological Seneca Ladles		9
Chapter II Eighteenth Century Seneca Archaeological Ladles		39
Chapter III Effigy Symbolism		51
Chapter IV Nineteenth Century Ethnological Ladles		61
Chapter V Changes in Effigy Motifs		91
Collections		97
Appendix		99
References Cited		111

FIGURES

Figure 1	Historic Seneca Sites Sequence	ix
Figure 2	Early Historic Seneca Villages, 1500–1687	x
Figure 3	Late Historic Seneca Villages, 1687–1820	xi
Figure 4	Bowl width-to-length ratio illustration	7
Figure 5	Angle of handle-to-bowl illustration	8

PLATES

Plate 1	Ladles and bowl from the Warren Site	13
Plate 2	Bowl effigy from the Warren Site	15
Plate 3	Ladle effigy from the Warren Site	17
Plate 4	Ladle effigy from the Warren Site	19
Plate 5	Ladle effigy from the Steele Site	21
Plate 6	Ladle from the Steele Site	23
Plate 7	Figurine from the Marsh Site	25
Plate 8	Ladle from the Marsh Site	27

Plate 8a	Detail of Plate 8	29
Plate 9	Ladle from the Boughton Hill Site	31
Plate 10	Ladle effigy from the Markham Site	33
Plate 11	Ladle effigy from the Power House Site	35
Plate 12	Ladle effigy from the Dann Site	37
Plate 13	Ladle effigy from the Snyder-McClure Site	43
Plate 14	Ladle from the Snyder-McClure Site	45
Plate 15	Ladle from the Honeoye-Morrow Site	47
Plate 16	Ladle from the Big Tree Site	49
Plate 17	Seneca ladle, Tonawanda Reservation	71
Plate 18	Seneca ladle, Cattaraugus/Tonawanda Reservation	73
Plate 19	Seneca ladle	75
Plate 20	Seneca ladle effigy	77
Plate 21	Seneca ladle	79
Plate 22	Cayuga ladle, Grand River Reservation	81
Plate 23	Onondaga ladle	83
Plate 24	Mohawk ladle	85
Plate 25	Mohawk ladle	87
Plate 26	Iroquois ladle effigy	89

CHARTS

Chart 1	17th Century Seneca Iroquois Archaeological Ladles	119
Chart 2	18th Century Seneca Iroquois Archaeological Ladles and one Ethnological Ladle	120
Chart 3	19th Century New York & Oklahoma Seneca Ethnological Ladles	121
Chart 4	19th Century Iroquois (Except Seneca) Ethnological Ladles	122
Chart 5	19th Century Ethnological Ladles According to Location When Collected, Mixed Tribal Groupings	123
Chart 6	19th Century Ethnological Ladles, Great Lakes Tribes	124
Chart 7	19th Century Algonquian Ethnological Ladles	125

PREFACE

The publication of *Research Records 15* represents what is expected to be a continuing series on Iroquois material culture, especially Seneca, as represented by the collections at the Rochester Museum & Science Center (RMSC) and, on occasion, in museums in Europe and North America.

The background research for *Aspects of Change in Seneca Iroquois Ladles A.D. 1600-1900* was initially conducted through an Arthur C. Parker Research Fellowship awarded to Betty Coit Prisch in 1978. Her subsequent investigations have provided the RMSC with a broad range of data on Iroquois ladles with both archaeological and ethnological provenience. The drawings by Gene Mackay were also the result of a similar fellowship in archaeological illustration designed to cover the entire range of Iroquois artistic expression. At the core of the study are the RMSC Seneca Iroquois collections that are part of a nearly unbroken Genesee region continuum acquired by the RMSC staff, members of the Lewis Henry Morgan Chapter of the New York State Archeological Association, and the Rock Foundation Inc.

Financial assistance for the publication of *Research Records 15* is gratefully acknowledged through the Arthur C. Parker Fund for Iroquois Research supported by the Rock Foundation Inc. The editor would like to express his thanks to Charles F. Wray, RMSC Research Fellow, and to Barbara Koenig, RMSC Public Information Officer, for their helpful suggestions; to Ann Prichard for her efforts as associate editor; to Richard Rose in his capacity as Curator of Anthropology and to Dana Stolka for her help in proofreading the manuscript. Finally, having both Betty Prisch and Gene Mackay in close proximity during the project has provided a constant stimulation which has greatly facilitated the research.

Aspects of Change in Seneca Iroquois Ladles A.D. 1600-1900 opens the way for further research on hitherto unavailable collections of Iroquois artifacts and original field notes, now at the RMSC. It also provides a frame of reference for increased communication among scholars.

Charles F. Hayes III
Research Director
Rochester Museum & Science Center

ACKNOWLEDGMENTS

Grateful acknowledgment is made to the Arthur C. Parker Fund for Iroquois Research for a Research Fellowship and grant in support of this study. Special appreciation is offered to my colleague, George R. Hamell, for his interest and guidance, and to Charles F. Hayes III, Charles F. Wray and Donald Cameron for their counsel and encouragement.

I am also grateful for their time and assistance to the following people: Stanley A. Freed, American Museum of Natural History, New York, New York; Jonathan C. H. King and Peter Gibbs, The British Museum, London, England; Raymond J. Hughes, Buffalo and Erie County Historical Society, Buffalo, New York; Virginia L. Cummings, Buffalo Museum of Science, Buffalo, New York; Phyllis Rabineau, Field Museum of Natural History, Chicago, Illinois; Thomas A. Breslin and Beverly A. George, Letchworth State Park Museum, Castile, New York; Nancy O. Lurie, Milwaukee Public Museum, Milwaukee, Wisconsin; James G. E. Smith, Museum of the American Indian—Heye Foundation, New York, New York; Anne Fardoulis and Marie France Fauvet-Berthelot, Musee de l'Homme, Paris, France; Horst Hartmann, Museum fuer Voelkerkunde, Berlin, West Germany; Rolf Gilberg, Nationalmuseet, Copenhagen, Denmark; Judy Hall, National Museum of Man, Ottawa, Canada; Charles Gillette, New York State Museum, Albany, New York; Sally Bond, Peabody Museum of Archaeology & Ethnology, Harvard University, Cambridge, Massachusetts; Joan E. Cohen, Peabody Museum of Natural History, Yale University, New Haven, Connecticut; John Richard Grimes, Peabody Museum of Salem, Salem, Massachusetts; and Lynne Williamson, Pitt Rivers Museum, Oxford University, Oxford, England.

Many busy scholars have taken time to share their expertise with me, and their help is sincerely appreciated. None of the above bears any responsibility for errors which remain solely mine.

Betty Coit Prisch

HISTORIC SENECA SITES SEQUENCE

Western	Eastern	Approx. Dates	Notes
(Little Beard)		1775-1820*	
Big Tree		1779-1820*	
Canawaugus		1775-1820*	Reservation Era
(Caneadea)		1775-1820*	
Fall Brook		1745-1775*	
Honeoye	Kendaia	1745-1775*	
(Conesus)	(Kanadesaga)	1745-1779*	Sullivan Expedition in 1779
(Avon Bridge)		1745-1779*	
(Canandaigua)	(Kashong)	1745-1779*	
Huntoon	Townley-Read	1710-1745*	
Snyder-McClure	White Springs	1687-1710	Denonville Expedition in 1687
Rochester Junction	Boughton Hill	1675-1687	
(Kirkwood)	Beale	1670-1687	
Dann		1660-1675	Susquehannock Defeat in 1675
	Marsh	1650-1670	
Power House		1645-1660	First Jesuit Mission in 1656
	Steele	1635-1650	Huron Wars 1642-1649
(Lima)		1625-1645	
(Bosley Mills)		1625-1645	
	Warren	1615-1635	
	(Cornish)	1615-1635	
(Dutch Hollow)		1600-1625	
(Feugle)		1600-1625	First Direct Trade
	Factory Hollow	1590-1615	
Cameron		1575-1600	
	Tram	1565-1590	Indirect Trade Era
Adams		1550-1575	European Epidemics First Trade Goods Formation of League
	(Richmond Mills)	1500-1550	Transition to Historic
	(Belcher)	1500-1550	End of Prehistoric Era

Figure 1. Listing the historic Seneca village sites sequence. (Sites listed in parentheses yielded no ladle data.) Dates followed by asterisks were revised by C. Wray, 1982. Other dates, Wray 1973.

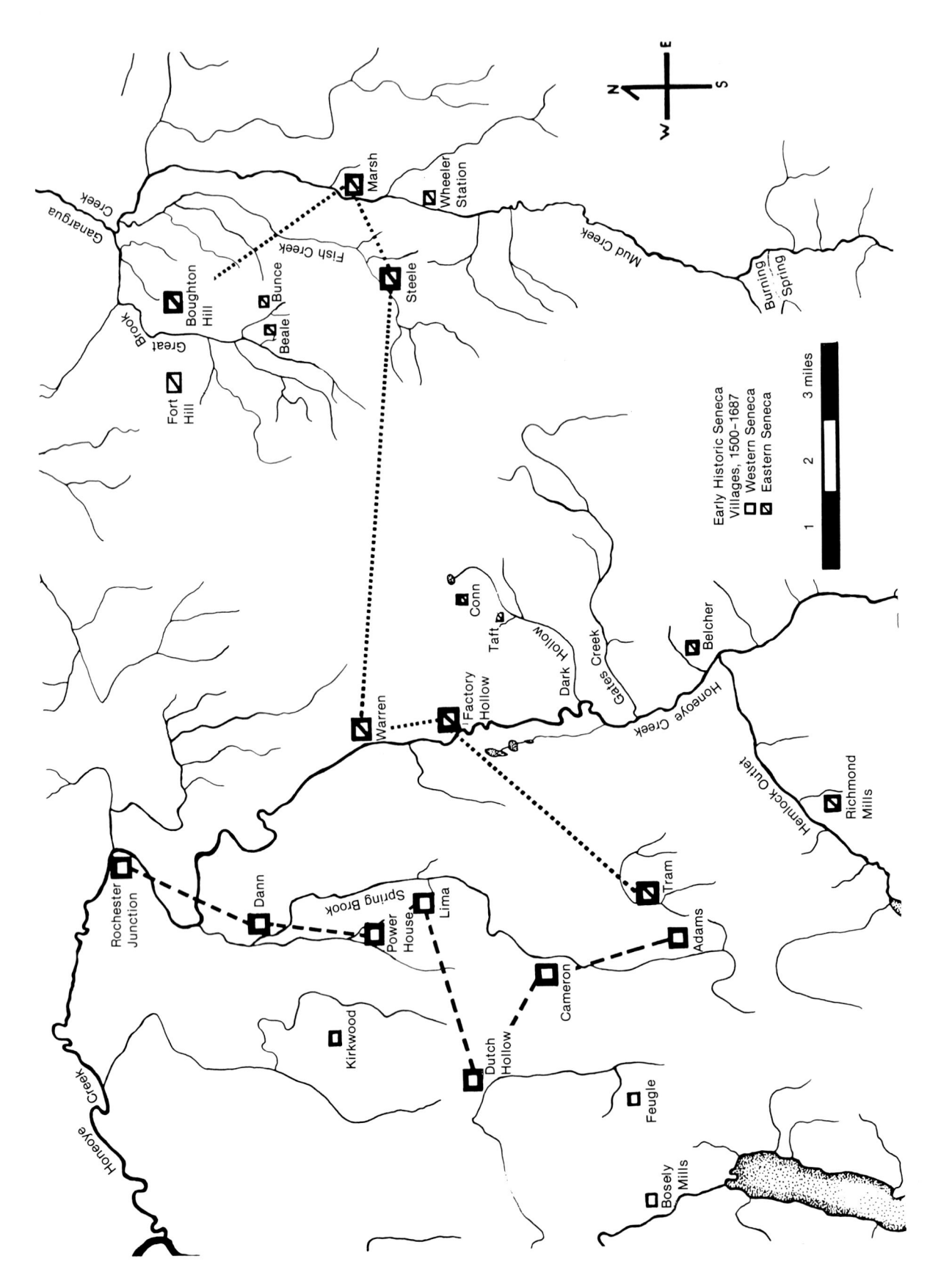

Figure 2. Early Historic Seneca Villages, 1500–1687 (Wray 1973)

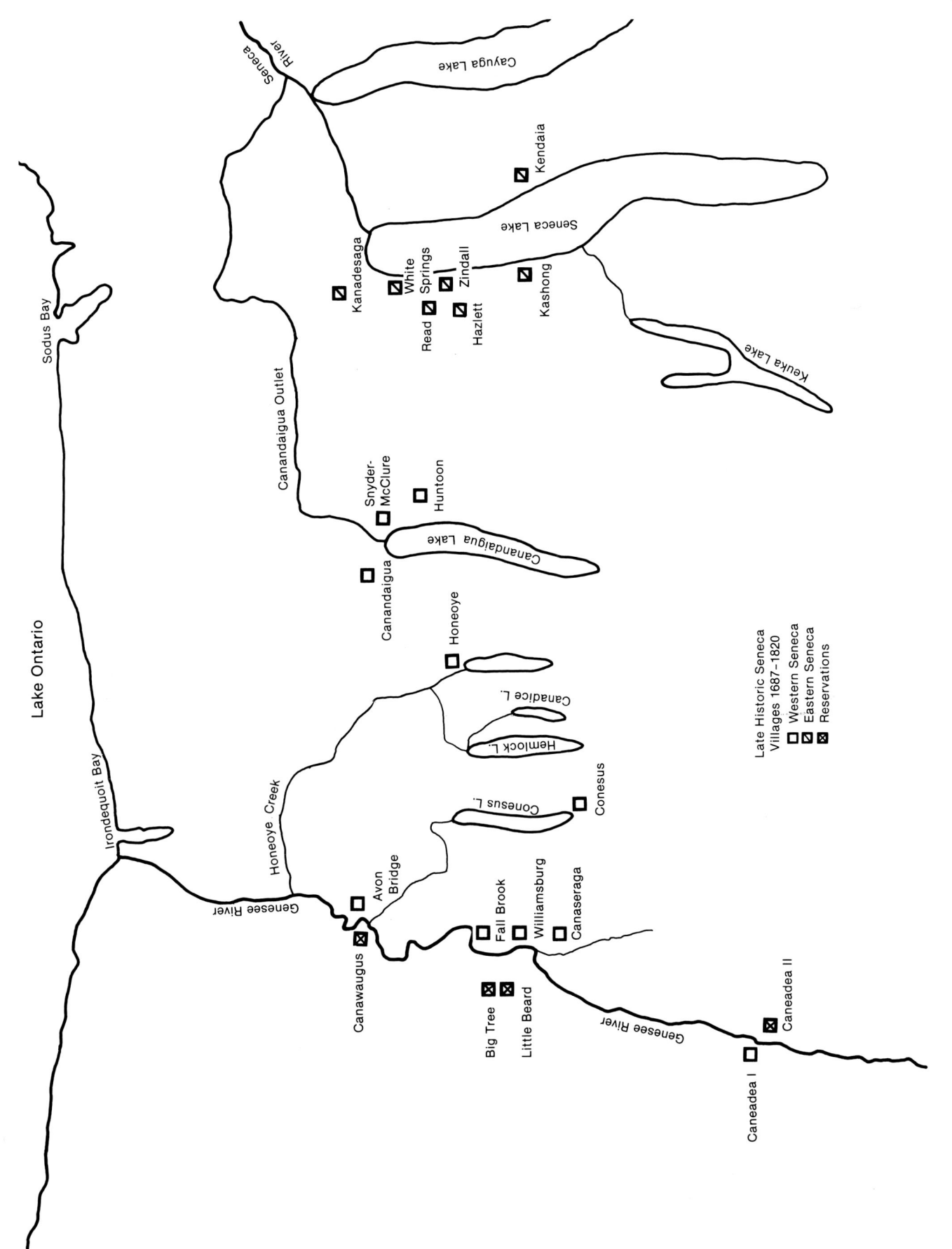

Figure 3. Late Historic Seneca Villages, 1687–1820 (Wray 1982)

GENERAL INTRODUCTION

The archaeologically preserved ladles of wood, antler and shell in the Wray-Cameron collection at the Rochester Museum & Science Center, all of which were excavated from New York Seneca Iroquois sites of the sixteenth to the nineteenth century, are part of the most extensive collection of early historic Seneca Iroquois artifacts in the Northeastern United States.

The ladles from this collection and from other Rochester Museum (RMSC) collections form the basis for the present study, which was undertaken to determine whether an analysis of changes in the morphology of the ladles and changes in their effigy finials could make a contribution to the ethnohistory of the Seneca.

Between c.A.D. 1600 and c.A.D. 1900, the data reveal systematic changes in the morphology and in the effigy carving of the ladles. It is suggested that the changes observed in the form of the ladles were adaptive functionally. Further, it is proposed that the effigies carved by the Seneca on their ladles were selected because they symbolized concepts of central importance to them, and that changes observed in these Seneca effigy subject matters between 1600 and 1900 were not arbitrary. Rather, the changes were responses to societal changes and may be understood in terms of events and conditions in the world of the Seneca people.

STUDY SAMPLE

Because they are the most numerous in the collections, Seneca ladles made of wood are the focus of this study. Some antler and shell ladles also are included. Several archaeologically preserved ladles from sites of other New York Iroquois groups of the seventeenth and eighteenth centuries are presented for comparison.

A sample of nineteenth century ethnologically preserved New York Seneca Iroquois ladles is contrasted with the seventeenth and eighteenth century archaeological ladles. Several ethnologically preserved eighteenth century ladles also are compared with their archaeological counterparts of the eighteenth century.

Data also were obtained on nineteenth century ethnological ladles of the neighboring northern and eastern Algonquians and tribes of the western Great Lakes, all peoples who had direct contact with the Seneca Iroquois as traders, as refugees, or as captives who were adopted and assimilated into the Seneca nation.

A total of 701 ladles is included in this study. Since the conclusion of the original work, a number of additional ladles has become available, so that a supplementary paper is projected.

Despite the sample size, it undoubtedly contains biases due to differential preservation, disposal and collection. A similar caveat is appropriate regarding the sample of ethnologically preserved ladles. For example, the existence of a ladle dated 1791 which was collected in the twentieth century from Cattaraugus Reservation raises the question of whether archaeologically preserved ladles truly represent examples from the century of the site in which they were found. In the same way, a ladle which was excavated from a sixteenth or seventeenth century site may have been an heirloom from an earlier century, creating an additional, unmeasurable bias in the study sample.

Because woodenware was standard among European households in the seventeenth and eighteenth centuries, data also may be skewed because European settlers did not see the native-made ladles as a novelty, and thus did not collect them in any significant numbers until the nineteenth century.

Further, some effigies included in the analysis may not be ladle finials, but rather may represent smoking pipe effigies (especially those inlaid with brass), figurines or maskettes.

The study does not purport to be exhaustive of collections of ladles of the Northeastern Woodlands Indians. Many important collections were not examined, primarily due to time limitations. It is proposed that the data presented be viewed as a springboard for additional study. A list of the ladle collections included in the study can be found in the Appendix.

Betty Coit Prisch is associate curator of anthropology at Rochester Museum and Science Center, Rochester, NY 14603

HISTORICAL PERSPECTIVE ON THE SENECA

The historic Seneca Iroquois were the most westerly and most populous member of the five nation (later six) League of the Iroquois, which probably was formed in the late sixteenth century by the Seneca, Cayuga, Onondaga, Oneida and Mohawk nations living in what became New York State (Fenton 1961:271). The Iroquoian-speaking League or Confederacy was surrounded by Algonquian-speaking native peoples (Fenton 1940). As "Keepers of the Western Door" of the Confederacy, the Seneca were on a frontier and shared cultural traits with their Algonquian neighbors as well as with the Canadian Huron Iroquoians. Later, the Seneca absorbed southeastern patterns from Siouan, Muskhogean and southern Iroquoian-speaking captives (Fenton 1940:167).

The seventeenth century Seneca were known to be composed of many tribes, 11, according to the Jesuits (JR 1656:265). Speaking of the Iroquois in general, the Jesuits remarked ". . . they have no neighbors to fight because they have subjugated all of them, [so] they go to seek new enemies in other countries" (JR 1656:263). A stimulus for the Seneca's expansionist policy was their desire to acquire fur pelts to trade with the Europeans for guns, metal knives, kettles, cloth, beads and other products which once were luxuries but which soon became necessities.

When European trade goods first began to filter into western New York early in the sixteenth century, the Seneca were living in two large and two associated small villages near present day Livonia, New York. They were organized into clans; inheritance passed through the maternal lineage.

The Seneca were a horticultural people, growing extensive acreage of corn, beans and squash. Women performed the every-day agricultural work, while men hunted and traded.

The use of eating ladles or spoons, common not only among the historic Seneca, but also throughout the historic Woodlands, probably stemmed from a considerable reliance on vegetable and meat soups. The well-documented dependence of the historic Iroquois upon what they termed their "life supporters," corn, beans and squash, is a central theme around which many ceremonial activities coalesce.

Eating ladles also are documented prehistorically by the archaeological record. The large earthenware pots recovered from prehistoric Seneca and other Iroquois sites also would seem to indicate a dietary base of boiled soup or stew.

SHELL UTENSILS

The use of shells as eating utensils is widely reported among natives of the historic period. Densmore records the use of clam shells as spoons by the twentieth century Chippewa (Densmore 1929:41). Paraphernalia for the Oklahoma Delaware Big House rite includes mussel shell spoons (Speck 1937:opp. p. 26). In the re-creation ceremony of the Menominee Medicine Lodge, Manabus, the culture hero, is given medicine from a "clam" shell (Skinner 1920a:45).

Pastorius (1912:384) observed eighteenth century Indians of Pennsylvania using mussel shells for spoons. Beverley illustrates a Virginia Indian couple of the eighteenth century eating with their fingers from a platter of food. Beside them is shown "a cockleshell, which they sometimes use instead of a spoon" (Beverley 1722:182–183). Waring and Holder (1945:13, 14) summarize the occurrence of engraved conch shell bowls among the artifacts of the Southeastern Ceremonial Complex which was documented in attenuated form as late as mid-eighteenth century among the Creek Muskhogeans, but which is known archaeologically as early as c.A.D. 950 (Willey 1966:304–306).

The use of shells as spoons in the seventeenth century is documented by Father Allouez who observed, on a visit to Wisconsin in 1669–1670, that "the savages of this region are more than usually barbarous; they are without ingenuity, and do not know how to make even a bark dish or ladle; they commonly use shells" (Scribner 1917:146).

Miniter suggests that the natives of North America taught the early European colonists "the art of sticking a clam shell into a split stick and calling it a spoon (deJonge 1973:504).

The protohistoric Seneca Iroquois Cameron Site (Hne 29-1) produced both unmodified clam shells and wooden ladles. The Cameron Site, which dates c.A.D. 1575–1600, also yielded a perforated, spoon-shaped shell artifact possibly used as a spoon and/or pendant. Unmodified clam shells also were recovered from the historic Seneca Dutch Hollow and Cornish Sites, c.A.D. 1600–1625 and c.A.D. 1615–1635, respectively.

The presence of unmodified fresh water mussel shells in earthenware pots is a trait of the protohistoric Late Woodland Seneca Iroquois, notably from the c.A.D. 1550 Adams Site (Hne 30-3) and the c.A.D. 1575–1600 Cameron Site (Hne 29-1) (Wray & Cameron: n.d.).

Shells used as eating utensils also are known prehistorically. Shells which have been modified to provide a handle as well as shells carved with effigies have been found in burial mounds of the Early Woodland Adena Culture in Tennessee, Kentucky, Ohio and Arkansas (Holmes 1883:199-200, Pl. XXIV).

Shells which have been cut and/or ground into a pear shape with a perforation at the smaller end have been recovered from the Late Archaic Stage Frontenac Island Site (Ritchie 1945:110, Pl. 12, figs. 36,38,39). These may have been used as spoons and/or pendants. They are similar to the Cameron Site perforated shell spoon/pendant.

Unmodified clam shells are known from the Late Archaic Stage Lamoka Lake Site (Hpt 1-3) (Ritchie 1932) and the Cole Gravel Pit Site (Hne 17-1) (Hayes and Bergs 1969).

ANTLER UTENSILS

Eating utensils carved of antler also are known in the historic period. Several antler ladles have been recovered from historic New York Seneca Iroquois sites. These will be discussed in the body of the study along with the wooden ladles.

An antler ladle (RMSC AR 17608) is known from the protohistoric Iroquoian Green Lake Site (Buf 1-4), Erie County, New York. Bone cups or ladles are known from the Owasco Sackett Site (Can 1-1) in Ontario County, New York (Ritchie 1936). Doubtless, there are other examples.

In the RMSC's collections, the earliest known examples of prehistoric antler ladles were recovered from the Late Archaic Stage Frontenac Island Site (Aub 4-3) in Central New York (Ritchie 1945; 1965). One of these utensils is non-effigy; the other is described as "fragmentary antler spoon with effigy top . . ." (Ritchie 1945:109, Pl. 11, fig. 26). This effigy is difficult to interpret; it may be the head of a bird.

WOODEN UTENSILS

The historic Seneca Iroquois used wood extensively in the artifacts of their material culture. Wood fueled the cooking and heating fires, and wood and bark were used as house building materials. Wooden bows and arrows were made for the hunt and wooden javelins and lacrosse sticks for sport. Wood also was carved into bowls and ladles for household use.

Although wooden artifacts probably represented a large portion of the material culture of the prehistoric Northeastern Woodlands Native Americans (judging from comparison with early historic accounts), a negligible amount has been preserved from archaeological settings. Wooden artifacts ordinarily are not recovered from prehistoric sites in the Northeastern United States due to soil conditions which promote decay. Although it is assumed that ladles of wood were used prehistorically, none which dates from pre-contact sites has been identified in the course of this study.

Other artifacts of wood have been found at prehistoric sites, preserved by unusual conditions such as total water immersion, extreme aridity or charring. Many North American examples are known: some are from such sites as Key Marco, Florida; Ozette, Washington; Roebuck, Ontario, Canada; and Chenango and Jefferson Counties, New York. At the Roebuck Site (a thirteenth century Iroquoian village), Wintemberg excavated pieces of worked wood from areas where middens extended into a swamp (MacDonald 1977:3). At the coastal Ozette Site, a preform for two wooden bowls was discovered (MacDonald 1977:2). The Key Marco Site yielded "cups, bowls, trays and mortars of wood" (Cushman 1896, quoted in Fundaburk 1957:96). In Chenango County, New York, a dugout canoe was dredged from "dense muck and clay" saturated with water (Whitney 1974:1). Beauchamp illustrates charred wood artifacts from Jefferson County, New York (Beauchamp 1905:160). It seems, therefore, more than probable that wooden ladles were in use in at least some of the prehistoric period.

The abrupt appearance of the wooden ladle in the archaeological record of New York State is due primarily to the introduction of brass trade kettles and their subsequent burial association with ladles. The anti-bacterial action of the copper salts leached from the kettles by ground water preserved organic materials in contact with them. Predictably, wooden pipes which have survived from this period have been inlaid decoratively with brass or have had functional brass bowl liners.

As previously noted, the wooden eating ladle appears to be common to all the historic tribes of the Eastern Woodlands of North America, and among native-made artifacts, it probably has had one of the longest histories of continuous use.

M. R. Harrington found that the manufacture of wooden spoons and bowls was one of the few native crafts still being practiced among the Canadian Delaware Algonquians of the early twentieth century (Harrington 1908:408). Similarly, in 1915, Frank G. Speck reported the continuing use of wooden spoons among the Mohegan Algonquians of Connecticut (Speck 1915:10). Among the Iroquois of the Six Nations Reserve, wooden ladles still are being carved in the fourth quarter of the twentieth century.

The New York Iroquois say that "food tastes much better when eaten from [a wooden ladle] and those who have not used them for some years express a longing to employ them again, recalling with evident pleasure the days when they ate from an *atog'washa*" (Parker 1910:56-57).

Although the material possessions of the early historic Native Americans appear to have been meager, the wooden bowl and ladle were omnipresent. Of the Seneca Iroquois, Mary Jemison noted ". . . our cooking and eating utensils consisted of a hommany [sic] block and pestle, a small kettle, a knife or two, and a few vessels of bark or wood" (Seaver 1967:47). Similarly, when Halliday Jackson settled some Quakers into houses purchased from the Seneca Iroquois, the Indians moved out "their goods and chattels, which . . . consisted chiefly in homony [sic] blocks and pounders, a brass kettle or two, some wooden bowls and ladles . . ." (Jackson 1830:31-32).

The eating bowl and ladle seem to have been owned individually. When the Mohawk Iroquois travelled, they took with them "some of their maize, a kettle, a wooden bowl, and a spoon; these they pack up and hang on their backs (Megapolensis 1644:174). After James Smith's capture and adoption into the Caughnewago Iroquois nation, he was given his own bowl and wooden spoon (Smith 1799:11). Guests at feasts of the Siouan Iowa each brought a bowl and spoon, and, after feasting, "all pack up their bowls and spoons and pass out dancing" (Skinner 1920a:204).

Many observers of seventeenth, eighteenth, and nineteenth century Native Americans focused on the large size of the ladles.

When Father Gabriel Sagard in 1632 was served corn meal mush by the Huron, the wooden spoon with which he ate was "as big as a small dish or saucer" (Wrong 1939:72).

Beverley noted (1722:154) that the ladles used by the eighteenth century Virginia Indians

> do generally hold half a pint; and they laugh at the English for using small ones, which they must be forc'd to carry so often to their Mouths, that their Arms are in Danger of being tir'd, before their Belly.

An Iroquois chief who in 1745 invited visiting Moravian missionaries to breakfast served tea and Indian bread, and "the tea cups were a very large spoon and a wooden bowl" (Beauchamp 1916:65).

Captured by Indians while travelling in 1791, Patrick Campbell was allowed to share with his captors some rum which they drank "out of a large wooden spoon that would hold a pint" (Campbell 1978:np).

The Rev. William Beauchamp, calling upon an Onondaga Iroquois friend in 1895, "surprised him at his meal. His spoon was as large as a wooden butter ladle, and his bean soup disappeared with corresponding rapidity" (Beauchamp 1895:216).

In summary, it is clear that among the Seneca Iroquois of New York of the period c.A.D. 1600–c.A.D. 1900 and later, the wooden eating ladle was in common use as evidenced by its archaeological recovery from sixteenth through eighteenth century sites and its ethnological preservation from nineteenth century reservations.

EFFIGY CARVING

In view of the large number of effigy ladles encountered in the course of this research, it is remarkable that references to effigies in the literature are rare. Antoine Denis Raudot wrote in 1709 of the Indians of the Upper Great Lakes that "the men make dishes of knots of wood and spoons on which they carve the figure of some animal" (Kinietz 1940:351). Speck says the Yuchi (Muskhogean? Swanton 1946) ladles have an "ownership mark (on the handle) consisting of a few scratches or incisions" (Speck 1909:42).

Most recorders were content to describe shape, as did Louis Phillippe when he visited the Cherokee Iroquoians in 1797, reporting that their wood spoons were "more pointed and more triangular" than European spoons (Sturtevant 1979:199).

In the middle of the nineteenth century, Lewis Henry Morgan could write:

> Their wooden implements were often elaborately carved. Those upon which the most labor was expended were the ladles, *Ah-do-quä'-sä,* of various sizes, used for eating hommony [sic] and soup. They were their substitute for the spoon, and hence every Indian family was supplied by a number. The end of the handle was usually surmounted with the figure of an animal, as a squirrel, a hawk, or a beaver, some of them with a human figure in a sitting position, others with a group of such figures in various attitudes, as those of wrestling or embracing. These figures are carved with considerable skill and correctness of proportion (Morgan 1962:383).

Nevertheless, an effigy carving tradition of considerable antiquity exists in the Eastern Woodlands. As was noted above, the Late Archaic Stage site of Frontenac Island produced a shell spoon with an effigy. Another effigy from this site is carved on an antler comb. The comb is surmounted by a symmetrical carving of two facing birds, beaks and breasts touching (RMSC AR 36032).

Also, as previously noted, shell effigy spoons are known from the burial mounds of the Early Woodland Adena Culture in Tennessee, Kentucky, Ohio and Arkansas (Holmes 1883:199-200).

Stylized human effigies on smoking pipes and earthenware pots are known from the Late Woodland Stage Owasco Culture, and variations on these persist into the historic period.

In summary, an effigy carving tradition in wood is associated with the historic natives of the Northeastern Woodlands, expressed in carved wooden masks, combs, smoking pipes, figurines and, particularly among the New York Seneca Iroquois, effigy carvings of human, animal or bird which, typically, ornamented the handles of their wooden ladles.

CARVING TECHNOLOGY

The technology of Native American wood carving as described by Father Louis Hennepin in the seventeenth century probably is equally applicable to the prehistoric period. Hennepin reports:

> "when the Savages are about to make Wooden Dishes, Porringers or Spoons, they form the Wood to their purpose with their Stone Hatchets, make it hollow with their Coles out of the Fire and scrape them afterward with Beavers Teeth for to polish them" (Hennepin 1903:103).

Hafted beaver incisors, "presumed wood-carving tools," were recovered from Early Woodland Meadowood Phase sites in western and central New York (Ritchie 1965:190) and from prehistoric Middle Woodland Point Peninsula Cultures, similarly located (Ritchie 1965:204).

Henry Hudson witnessed early seventeenth century Native Americans using shells as knives for the skinning of animals (Jameson 1909:49), perhaps indicating use of shell knives for carving wood as well.

The Virginia Indians reported to visiting ship captains as early as 1584 that they were accustomed to making tools (possibly knives) from nails and spikes recovered from shipwrecks (Anon 1600:247-8 in Pratt 1976:29).

In the historic period the crooked knife became the preferred tool for wood carving, especially of concave surfaces of bowls and ladles. Ritzenthaler says the Indians' wood carving knife was the "farrier knife, introduced by the Europeans and used by the blacksmith in the shoeing of horses" (Ritzenthaler 1976:34). Hanson, however, distinguishes between the farrier's or hoof knife and the canoe or crooked knife, noting that the former is forged in a gradual curve throughout much of its length and ends in a sharp inward curve. The canoe knife is forged flat for most of its length, then bent upward in a regular curve. The Eskimo crooked knife is different from either of these. It has a very short blade (Hanson 1975:9).

Quimby records a list of trade goods taken into Illinois country by a fur trader in 1688 which included six dozen canoe knives (Quimby 1966:65). Lahontan in 1703 lists the crooked knife in his "Algonkin" dictionary as *Coutagan* (Lahontan 1703:739). Under its Cree name of *Mocotaugan,* the crooked knife is listed in the Hudson's Bay Company catalog of 1748 (Quimby 1966:66).

Maurer (1977:101) illustrates a crooked knife, believed to be Seneca c.A.D. 1850, which has a carved wooden haft of a human head and torso effigy. The blade is made of a straight razor. Many crooked knives were manufactured by blacksmiths and also by the Indians themselves, using a file or section of a knife (Hanson 1975:6).

Wood carving was men's work. Adriaen van der Donck, describing the natives of New Netherlands in a publication of 1650, reported "the men . . . make spoons, wooden bowls, bags, nets and other similar articles" (Jameson 1909:302). Lawson says of the southeastern Indians in general, ". . . those that are not extraordinary hunters make Bowls, Dishes and Spoons of Gum-Wood and the Tulip Tree" (Lawson 1709:208). Of the New England Indians, Gookin reported in 1792, "Their dishes, and spoons, and ladles, are made of wood, very smooth and artificial, and of a sort of wood not subject to split. These they make of several sizes . . . their dishes, pots, and spoons, are the manufacture of the men" (Gookin 1970:16).

Iroquois bowls often were carved from tree knots, but most Iroquois ladles were carved in one piece from a block of wood (Parker 1910:56). Traditionally, then, Iroquois men are the wood carvers.

EUROPEAN CARVING

The Europeans brought with them to the New World a wood carving tradition. In Europe, woodenware, called treen ware, was the common household kitchen and table equipment which in the seventeenth century only slowly was being displaced by pewter in more well-to-do homes (deJonge 1973:504). Early colonists, particularly those from Northern Europe, came from countries with plentiful wood supplies and a dairy industry which depended on wooden utensils for the production of butter and cheese. The colonists set forth for the New World with a full complement of woodenware, as it was durable, lightweight and easily replaced. Upon arrival, they found "the Indians unconsciously in fashion" with their wooden ladles and bowls (Miniter 1973:n.p.). The Europeans' familiarity with woodenware, as mentioned above, probably accounts for the scarcity of early historic Indian examples in European collections.

Wooden spoons of European manufacture are common in Old World collections. In addition to utilitarian dairying spoons and ladles, there are many examples which have handles carved in delicate filigree or ornamented with complex human and animal effigy carvings.

Although this study does not purport to make an exhaustive investigation of European wood carving traditions and related peasant crafts, it is important to note, however briefly, that there are some examples of cups, spoons and jewelry which show decorative techniques and forms similar in concept to North American native craft work. Consequently, it is sometimes difficult to determine the direction of flow in the diffusion of material cultural traits. Relevant examples are illustrated in Bossert 1977: Plates 2,12,17,21, *Peasant Art of Europe and Asia*.

METHODOLOGY

In addition to the steadily increasing influence of the Europeans, beginning around the second quarter of the sixteenth century, the Seneca may also have absorbed cultural traits from their tribal neighbors as they expanded their sphere of influence in search for furs to trade. As the wooden eating ladle is common to all the tribes of the Eastern Woodlands, and as the varying cultural influences probably affected its design in many unverifiable ways, this study, therefore, does not search for a pure Seneca type. Rather, it seeks to understand the changes that can be observed between c.A.D. 1600 and c.A.D. 1900 in the various attributes of the ladles. This is undertaken despite recognition of the unverifiable amount of bias in the sample. It is hoped that the size of the sample may mitigate this. The conclusions drawn from the data more properly may be termed suggestions which seem to be indicated by the data.

The attributes of the ladles which contributed to the data include:
1) total length
2) bowl width-to-length ratio (See Figure 4.)
3) angle of handle-to-bowl (See Figure 5.)
4) effigy

Total length did not show systematic change over the years. Ladles examined in this study ranged in length from 5 cm to more than 30 cm. The longer ladles probably represent examples of serving ladles rather than individual eating ladles. (The longer ladles sometimes are referred to in the literature as feast ladles.) A majority of the ladles in the study ranged in length between 10 cm and 20 cm and were probably individual eating ladles.

Bowl width-to-length ratio showed a trend over time from a circular bowl among the oldest ladles to a bowl which was more wide than long. Examples of both, however, can be found in seventeenth, eighteenth and nineteenth century ladles. A bowl which is longer than wide was rarely seen in the study sample. It is, however, the typical form of European metal spoons which evolved over these centuries from a fig shaped bowl to a long oval, egg shaped bowl (Rainwater 1976:34).

The angle of handle to bowl apparently changed from an angle of 180° (that is, handle and bowl being carved in the same plane) to an obtuse angle of about 130° to 140°. Caution was exercised in measuring the angles of the archaeological examples, as the probability of warping was recognized.

The effigy probably was the most significant physical attribute of the ladles. The percentage of effigy versus non-effigy ladles was calculated and was found to vary significantly from one native group to another.

The ladles in this study were classified as effigy, abstract effigy, or non-effigy. The abstract classification, which is admittedly subjective, was bestowed on those ladles whose handle finials could not be given a recognized name. These may be "pure abstractions," or they may only "appear so until one learns what is intended" (Fenton 1950:25). An interpretation of the symbolism of the effigies was attempted in an effort to learn what was intended.

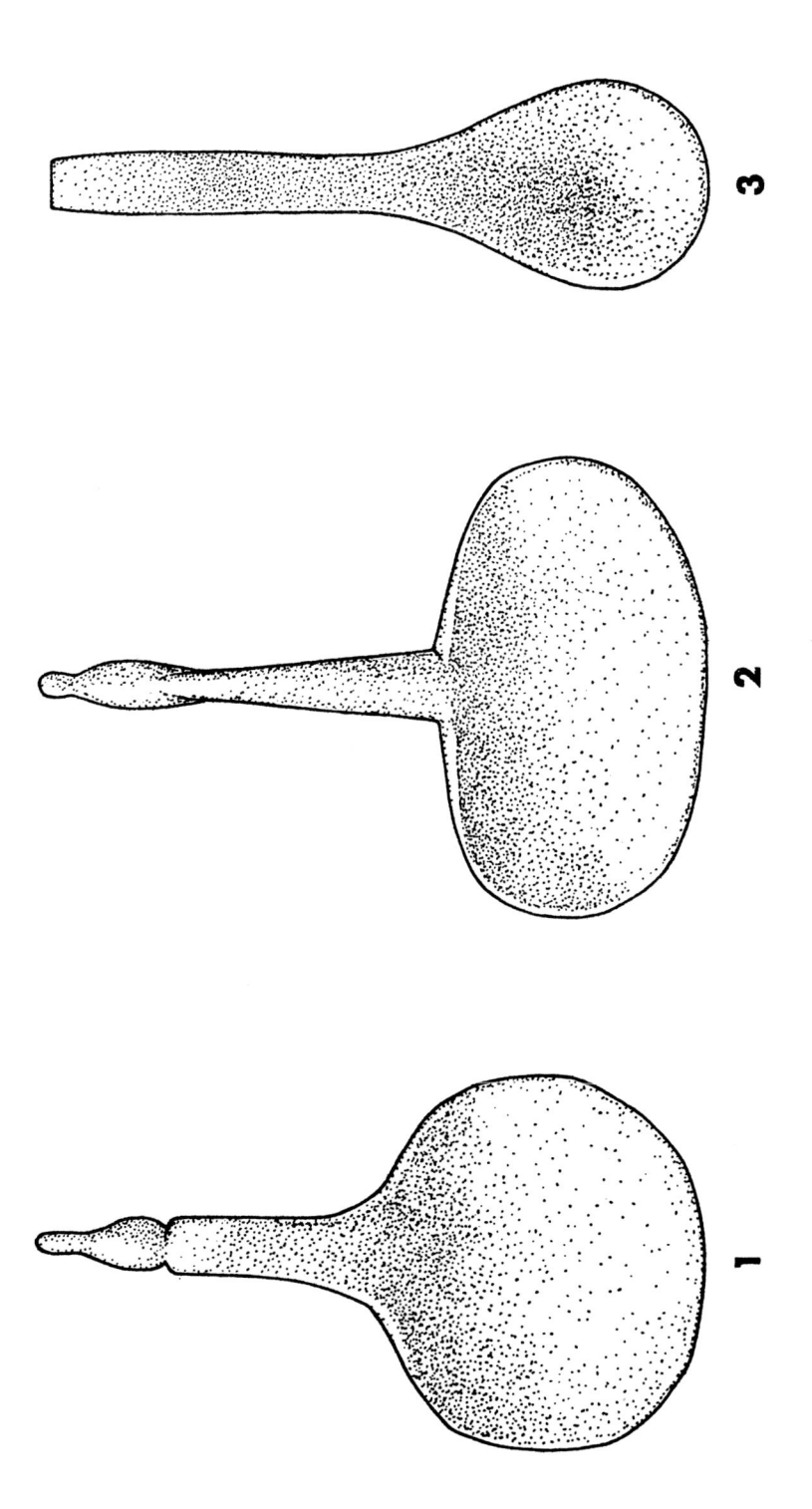

Figure 4. Illustrating the changing ratio of bowl width-to-length in Seneca ladles. Drawing by Gene Mackay.

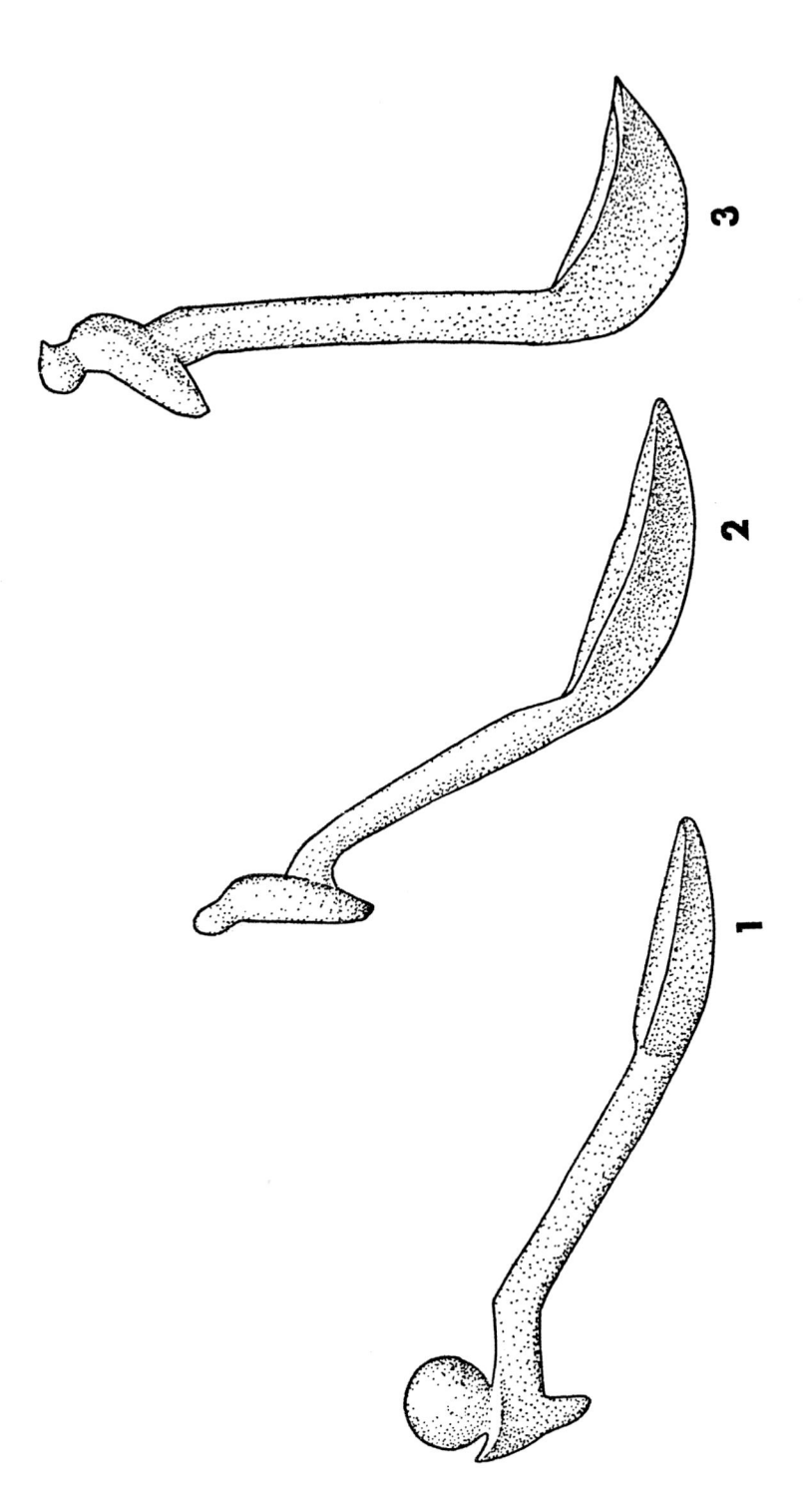

Figure 5. Illustrating the changing angle of handle-to-bowl in Seneca ladles. Drawing by Gene Mackay.

CHAPTER I

SEVENTEENTH CENTURY ARCHAEOLOGICAL SENECA LADLES

The parallel development of an eastern and a western sequence of historic sixteenth and seventeenth century Seneca Iroquois villages was alluded to earlier in this paper. The villages apparently were moved a few miles every 15 to 20 years after local game, soil and firewood were exhausted. The northward drift of these villages between c.A.D. 1500 and c.A.D. 1700 has been traced archaeologically (Wray and Schoff 1953; Wray 1973). The chart and map (after Wray 1973) illustrate the eastern and the western sequences. (See Figures 1 and 2.)

A summary follows of the ladles recovered from each village. A complete tabulation of ladles in the study can be found in the Appendix. Each ladle is numbered in the Appendix; references to ladles in the text are followed by the ladle number in parentheses.

EASTERN SEQUENCE

Factory Hollow Site

Of the eastern villages, the Factory Hollow Site (Hne 7-2) is the earliest in this study from which wooden ladles have been preserved. It is also representative of the period when brass trade kettles became common and began to be used as containers rather than as raw materials to cut up for decorative ornaments. The settlement at the Factory Hollow Site (c.A.D. 1590) was composed of people who had moved from the earlier (c.A.D. 1565–1590) Tram Site (Hne 6-4), which was the earliest in the sequence of eastern protohistoric Seneca villages. Of the seven Factory Hollow Site ladles, (6, 7, 8, 9) four have broken tops, one (3) is perforated at handle top, and another (4) is surmounted by a scalloped three-tiered pyramid. A final ladle (5) from the Factory Hollow Site has an animal effigy finial (probably bear, wolf or panther). The head is turned back over the shoulder.

Warren Site

From the c.A.D. 1620–1640 Warren Site (Hne 10-2), the next eastern village in the series, a wooden bowl with two wooden ladles nested inside was recovered. The bowl effigy (10) is of a human head, facing to the left. One ladle (12) effigy may be of a bear (the snout is blunted), or possibly a panther (there is some indication of a long tail). The other ladle effigy (11) is a hawk or eagle, recognizable by its hooked beak. (See Plates 1–4.)

Steele Site

In c.A.D. 1635, the eastern Seneca moved their village from the Warren Site to Fish Creek, where one antler and 13 wooden ladles were recovered from the Steele Site (Can 8-1). Two ladles (25, 26) have broken tops; one has a non-effigy, perforated top (24). Another (23) has a striated, tapered top; two (20, 21) terminate in what appear to be turtle head effigies; another (18) in a wolf or bear head effigy. One handle fragment (22) appears to have a snake winding around it. A human head (14), painted or stained light blue, is carved on another handle fragment. (See Plate 5.) A pair of stylized human heads face outward at the top of one handle (15). The first of the clenched-fist effigies (16) appears, and, on another ladle (17), a stylized clenched-fist representation is found. One ladle (19) bears an animal head effigy which resembles a horse. (Wentworth Greenhalgh's trip on horseback through the country of the Seneca did not occur until 1677, but it is likely that the Seneca had seen horses on visits to Fort Orange or Montreal and perhaps at Onondaga in the 1650s.)

The antler ladle (13) from the Steele Site has a carving, in profile, of an elderly female holding a cane. The spine is carved in jagged angles, as if to emphasize the vertebrae; the breast is pendulous, and some kind of headdress, or possibly hair cut in bangs, is represented. (See Plate 6.)

Marsh Site

The eastern Seneca moved c.A.D. 1650 to Mud Creek on the Marsh Site (Can 7-1), from which there are 17 wooden and two antler ladles. Of the wooden ladles, (72, 73, 74) three have broken tops, and one (71) is perforated through the handle. Two fragmentary handles display a hawk or eagle effigy (64) and a turtle head effigy (69). A stylized human head and torso (58) and a human female head and torso (59) surmount two ladles. An effigy (56) of a pregnant human female may have been a ladle handle finial. (See Plate 7.) A small ladle (70) displays an hourglass-shaped finial. A pair of bears (65), shown side by side facing away from the bowl, is carved on one ladle. (See Plates 8, 8a.) A panther (68) with head turned back over the shoulder is carved on another. One ladle finial (61) is carved as a strapped cylinder; it is conceivably a stylized, clenched-fist effigy. Another (60) with broken top also may have terminated in a clenched fist. Another broken effigy (62) is recognizable as a reclining human figure.

Among the ladles recovered from the Marsh Site, four are in the collection of the Ontario County Historical Society in Canandaigua, New York. Of the two wooden ladles, one (67) has a bear head and the other (66) a wolf head effigy. Of the two antler ladles, one (57) has a human figure effigy somewhat reminiscent of the Steele Site antler ladle, but more fragmentary. The head is carved with a turban-like ridge around the head, following approximately the hairline. The other antler ladle (63) shows the full figure, in profile, of a long-billed, long-necked, long-legged water bird—possibly a heron, crane, or bittern.

One of the Marsh Site ladles (73) with a broken handle actually may be a small bowl, with a handle which looks more like a lug. A similar example, on exhibit in the Field Museum of Natural History in Chicago, is captioned "small wooden bowl with snake like head carved projection used in preparing medicine." This bowl is attributed to the Sauk & Fox Algonquians.

Boughton Hill Site

The village at the Boughton Hill Site (Can 2-2), also called Gannagaro State Historic Site, was the next to be established by the eastern Seneca (c.A.D. 1675). Boughton Hill and its satellite village at the Beale Site (Can 10-1) were destroyed in the expedition of the French and their Indian allies under the command of the French governor of Canada, the Marquis de Denonville, in 1687 (Hamell 1980a). From the Boughton Hill Site were recovered three wooden ladles (86, 87, 88) with broken tops; two handle fragments (80, 81) with turtle head effigies; a wolf head effigy (79); two (77, 78) with the full figure of a bear, and one (75) with a reclining human figure effigy. A unique shell ladle from the Boughton Site has a turtle in bas relief on the handle (82). (See Plate 9.) In the New York State Museum collection from Boughton Hill Site, there is a wooden ladle (76) with a nine-ringed grip, an antler spoon (83) with a bifurcated finial, plus two other wooden ladles (84, 85) with broken tops.

Markham Site

The Markham Site (Hne 13-1), which also is given a terminus date of 1687 (because of associated datable artifacts), yielded one ladle (51) with owl head effigy. (See Plate 10.)

WESTERN SEQUENCE

Cameron Site

In the western sequence of seventeenth century Seneca Iroquois villages, the earliest wooden ladles to be preserved date from the c.A.D. 1575-1600 Cameron Site (Hne 29-1). The Cameron Site is the second protohistoric village (after the Adams Site, Hne 5-4) in the western Seneca sequence. It yielded two ladles (1, 2) which are wooden, non-effigy and have an elongated pear shape. The handles flow into the bowl without a delineating break, similar to the antler spoon (RMSC AR 37093) from the Frontenac Island Site of the Archaic Stage.

Power House Site

The next ladle-producing site in the western Seneca sequence is the Power House Site (Hne 2-2), dating c.A.D. 1645-1660—leaving a 50-year gap for which there is no ladle data to compare with the eastern villages. Effigies on smoking pipes and combs at the intervening sites, however, appear to continue in the same tradition. One Power House Site ladle (31) is non-effigy, another (33) has a broken top. One (32) is carved with a scalloped edge. A handle fragment of another (28) is carved to represent an animal head, and the last (27) appears to be a handle fragment with a toggle head.

Also found at the Power House site is a wooden effigy (29) of a bear, carved naturalistically, in the round. (See Plate 11.) Fracture marks on its underfeet indicate it may once have been a ladle handle or bowl effigy. Around the bear's body are faint indentations, which may have been made by bindings.

An antler ladle (30) from the Power House Site shows a long-tailed otter (or possibly salamander) in profile, climbing up the side of the handle.

Dann Site

Of the 15 ladles from the next western Seneca village, the Dann Site (Hne 3-2) of c.A.D. 1660–1675, five (46–50) have broken tops. Two (43, 44) are non-effigy, and one handle fragment (42) is perforated. An effigy (41) of a turtle head (possibly a smoking pipe effigy), inlaid with brass, is presumed to be a handle fragment. Another handle fragment (34) is topped with a human head which shows a carved, turban-like ring around the head, following the hairline. (See Plate 12.)

A ringed cylinder effigy ladle (37) may be a stylized clenched fist (there are five rings). There also is a ringed grip (35) with seven rings and another ladle (36) with a strapped cylinder effigy. One ladle (40) has a beaver effigy on the handle. An antler ladle (38), although fractured at the top, has a recognizable reclining human figure effigy. (Other reclining human figure effigies were found on ladles from the Marsh and the Boughton Hill Sites and on an antler comb from the Power House Site.) Two additional Dann site ladles (total 17) are in the New York State Museum collection; one of antler (39) has a swimming beaver effigy, the other (45) has a broken top.

Rochester Junction Site

The western Seneca Rochester Junction Site (Hne 11-2), one of the villages burned in the 1687 Denonville expedition, is represented by four ladles. Two are of wood, one (55) with a broken top and one (53) with a hawk or eagle effigy. Two are of antler. The first is of a bear (54) in profile, standing on its hind legs, with paws raised (a wolf or bear effigy is incised on the back of this ladle). The other ladle (52) which is possibly related thematically to the Steele Site antler ladle, shows a standing human in profile with a fragment of what may be an arm thrust forward. The lower spine is carved in jagged angles; there is a protrusion from the top of the head which could represent a horn. It is possible that the figure was holding a cane; the condition of the ladle is poor.

SUMMARY OF ATTRIBUTES

In summary, of the 90 seventeenth century Seneca ladles in the study, there are:

    10 human head or figure
    5 clenched human fist
    3 reclining human figure
    12 bear or wolf
    1 pair of bears
    1 panther
    1 horse
    1 otter
    2 beaver
    3 hawk or eagle
    1 owl
    1 heron/crane/bittern
    7 turtle
    1 snake
    1 hourglass
    1 fork top
    7 abstract
    8 non-effigy
    24 top broken

In addition to the effigies, other attributes of the ladles were examined. These included total length, ratio of bowl width to bowl length, angle between bowl and handle, and handle shape.

There were no apparent meaningful distinctions for seventeenth century ladles in total length. The smallest ladles (a non-effigy example from the Cameron Site, a ladle with a mammal effigy, a stepped pyramid ladle from the Factory Hollow Site and a specimen with an hourglass effigy from the Marsh Site) were 6.2 cm, 7.5 cm, 6.0 cm and 8.0 cm in length, respectively. These ladles do not seem large enough to serve as eating utensils; it is unlikely that they were baby spoons. It is possible that they were used to prepare or administer medicine, but there is no proof of this. Nineteenth century ethnological spoons which are characterized as "medicine" spoons do not share any commonality of size, shape or design.

Seventeenth century ladles are characterized by a bowl width-to-length ratio approaching 1.0 on the average, indicating an almost circular bowl.

For archaeologically preserved ladles, the angle between bowl and handle may be distorted by warping, so that conclusions would be misleading.

The handles of seventeenth century ladles are straight, without a back hook. Most of the effigies are carved essentially two-dimensionally, in the same plane as the bowl or nearly so, similar in concept to the effigies on antler and bone combs. Few examples are carved in the round; outstanding among these are the Warren Site human, bear and hawk effigies; the pair of bears effigy and the pregnant human female effigy from the Marsh Site, and the Power House Site bear effigy which is possibly from a ladle or bowl.

OTHER IROQUOIS LADLES

The small sample of seventeenth century ladles from sites of other New York Iroquois tribes has attributes comparable to the seventeenth century Seneca ladles. A Cayuga ladle (114) from the Rogers Site (c.A.D. 1670–1700) has a straight, non-effigy, perforated handle. Two Onondaga ladles have human head effigies. Both show traces of head bands. One (113) is from the Jamesville Pen Site (c.A.D. 1680–1700) and the other (112) from the Jayne LaPoint Site (c.A.D. 1650). All of these ladles are in private collections.

NON-NEW YORK IROQUOIS LADLES

Non-New York Iroquois seventeenth century ladles include four from the Neutral Iroquoian Grimsby Site, Ontario, Canada (ROM collection). One (115) has a human head effigy, with headband; another (118) has a weasel effigy, and two (119, 120) are non-effigy. Skinner illustrates two antler ladles of the Canadian Neutral Iroquoians; one (117) is non-effigy, the other (116) has a wavy handle perhaps representing a snake (Skinner 1920b).

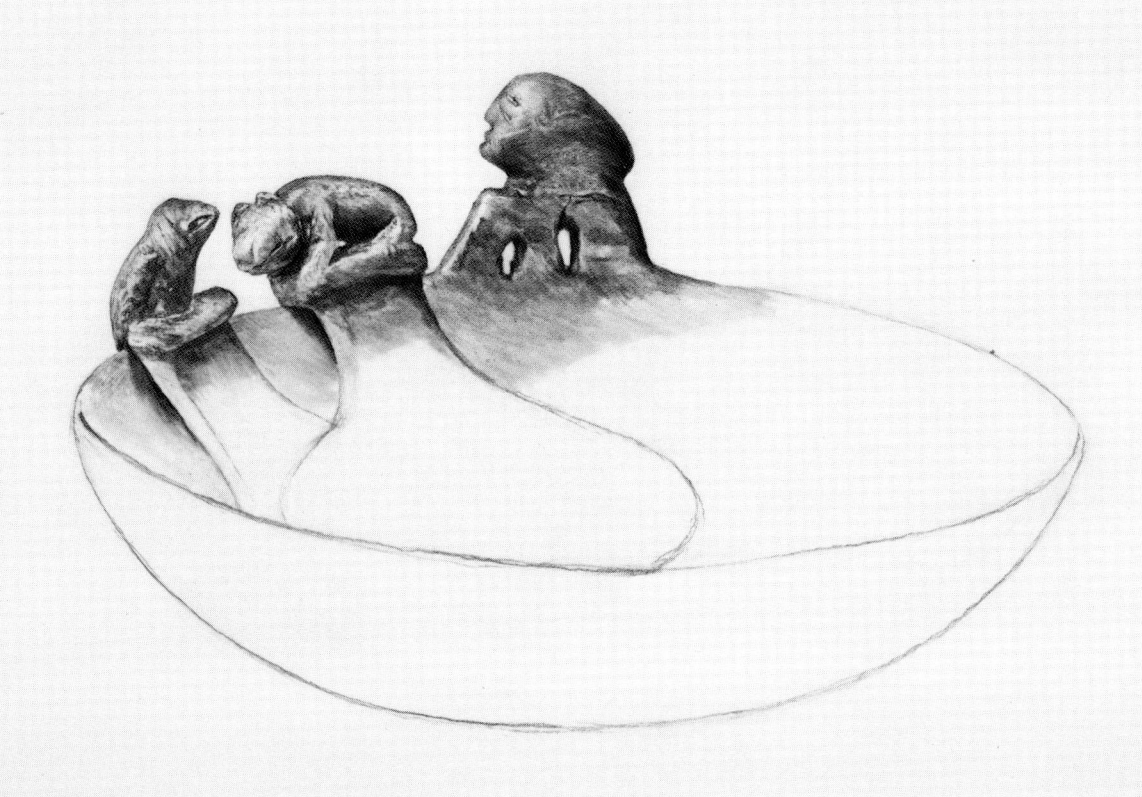

Plate 1. Ladles and bowl, wood. *Archaeological*. (Listed in Appendix as 10, 11, 12.) Warren Site (Hne 10-2). Seneca Iroquois, c.A.D. 1615-35. RMSC 40/89. About three-quarters actual size. Drawing by Gene Mackay. Note: Reconstruction of positions and forms of two effigy-decorated wooden ladles found resting within an effigy-decorated wooden bowl; all suffering some crushing and distortion during burial.

Plate 2. Bowl effigy, wood. *Archaeological*. (10) Warren Site (Hne 10-2). Seneca Iroquois, c.A.D. 1615-35. RMSC 40/89. About two and one-half times actual size. Drawing by Gene Mackay.

Plate 3. Ladle effigy, wood. *Archaeological.* **(12) Warren Site (Hne 10-2). Seneca Iroquois, c.A.D. 1615-35. RMSC 40/89. About two and one-half times actual size. Drawing by Gene Mackay.**

Plate 4. Ladle effigy, wood. *Archaeological*. (11) Warren Site (Hne 10-2). Seneca Iroquois, c.A.D. 1615-35. RMSC 40/89. About two and one-half times actual size. Drawing by Gene Mackay.

Plate 5. Ladle effigy, wood. *Archaeological*. (14) Steele Site (Can 8-1) Seneca Iroquois, c.A.D. 1635-50. RMSC 522/100. About two times actual size. Drawing by Gene Mackay. Note: The ladle effigy retains most of its original pale blue paint.

Plate 6. Ladle, antler. *Archaeological*. (13) Steele Site (Can 8-1). Seneca Iroquois, c.A.D. 1635-50. RMSC 5000/100. Actual size. Drawing by Gene Mackay.

Plate 7. Figurine, wood. *Archaeological.* (56) Marsh Site (Can 7-1). Seneca Iroquois, c.A.D. 1650-70. RMSC 807/99. About three times actual size. Drawing by Gene Mackay.

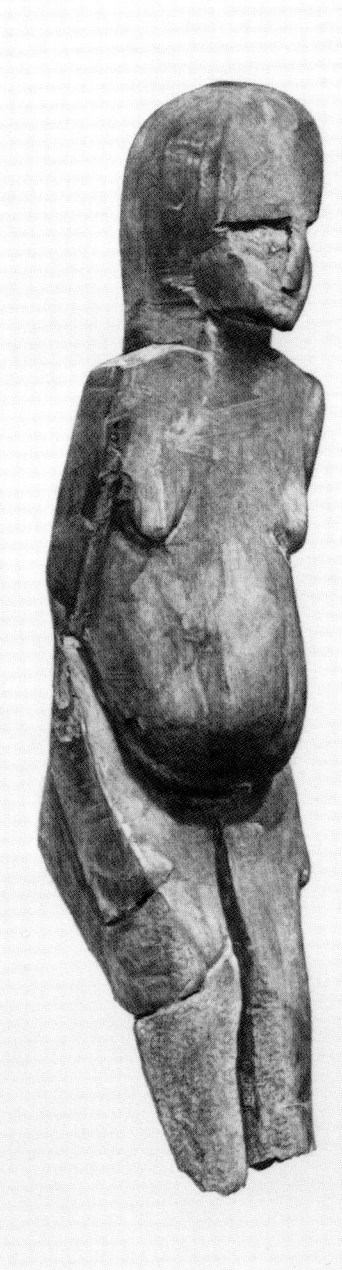

Plate 8. Ladle, wood. *Archaeological.* **(65) Marsh Site (Can 7-1). Seneca Iroquois, c.A.D. 1650-70. RMSC AR 18401. About one and one-quarter times actual size. Drawing by Gene Mackay.**

Plate 8a. (65) Detail of Plate 8.

Plate 9. Ladle, shell. *Archaeological.* **(82) Boughton Hill Site (Can 2-1) Seneca Iroquois, c.A.D. 1670-87. RMSC 2212/103. About one and one-half times actual size. Drawing by Gene Mackay.**

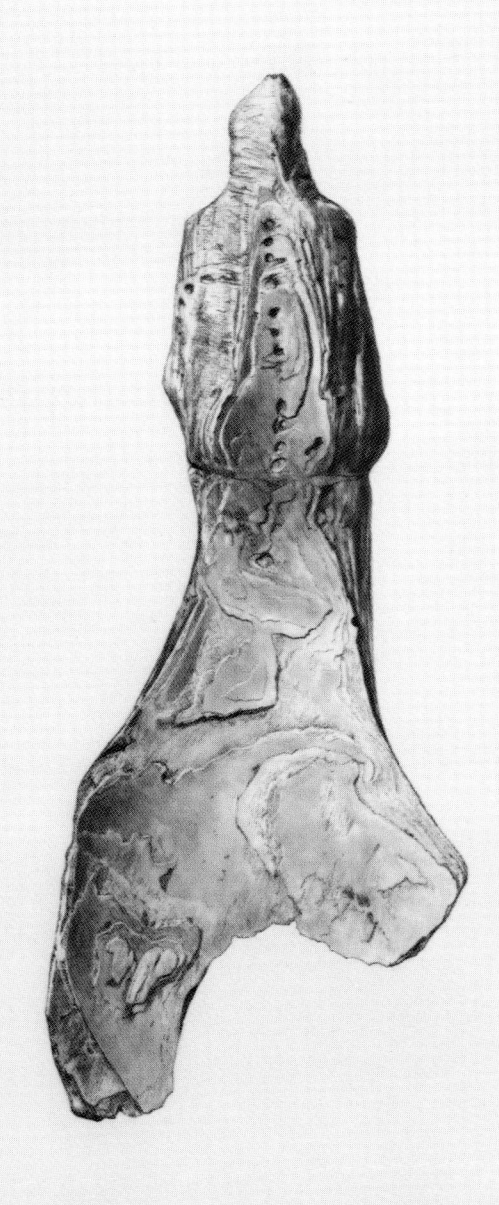

Plate 10. Ladle effigy, wood. *Archaeological.* (51) Markham Site (Hne 13-1). Seneca Iroquois, c.A.D. 1687. RMSC 145/T. About two times actual size. Drawing by Gene Mackay.

Plate 11. Ladle effigy, wood. *Archaeological.* (29) Power House Site (Hne 2-2). Seneca Iroquois, c.A.D. 1645-60. RMSC AR 42684. About two and one-half times actual size. Drawing by Gene Mackay. Note: The impressions of bindings on the surface of the effigy suggest that it had broken off during the ladle's use and subsequently was bound to the ladle's handle.

Plate 12. Ladle effigy, wood. *Archaeological.* **(34) Dann Site (Hne 3-2). Seneca Iroquois, c.A.D. 1660-75. RMSC 1123/28. About three times actual size. Drawing by Gene Mackay.**

CHAPTER II

EIGHTEENTH CENTURY SENECA ARCHAEOLOGICAL LADLES

The year 1687, marking the occasion of the French and Indian raid led by the Marquis de Denonville into Seneca Iroquois territory and the destruction of the four principal Seneca villages, effectively terminated a distinct period of Seneca history. Although some of the Seneca population reconverged into compact villages, by the second quarter of the eighteenth century the eastern and western lineages seem to break down, and it is more difficult to trace the movements of people as they spread out through the Genesee region.

There is some evidence, however, that the Snyder-McClure Site (located near Canandaigua, New York) received the population of the western village at the Rochester Junction Site (one of the villages destroyed as a result of the Denonville raid), that the Huntoon Site followed Snyder-McClure in the western sequence, and that the c.A.D. 1750–1779 Honeoye Site received the Huntoon Site and Snyder-McClure Site populations (Charles Wray, personal communication).

Snyder-McClure Site

The turn of the century Snyder-McClure Site (Plp 6-3), c.A.D. 1687–1710, has been included with the eighteenth century sites. Ladles from the Snyder-McClure Site show a continuity in traditional motifs. A reclining human figure ladle (91) has been preserved (see Plate 13), as well as one (92) with an animal head effigy. A brass-inlaid turtle effigy (93), which may be a smoking pipe fragment rather than a ladle effigy, was recovered. Another (95) features a perforated circle finial. A small horn ladle (94) displays a rayed hourglass finial, similar to the one from the Marsh Site. (See Plate 14.) In this ladle, the handle rises from the bowl at an angle of about 130°.

Huntoon Site

The c.A.D. 1710–1730 Huntoon Site (Plp 23-1) was separated from the Denonville Expedition by a generation. The Huntoon Site yielded a wood effigy fragment (100) with a human face carving coated with red pigment. (This may be a maskette of a False Face rather than a ladle handle.) A second Huntoon ladle (101) has a scalloped finial which conceivably could have been intended as a representation of human buttocks. The handle is warped, and it is not clear what was its original position. The grain of the wood, however, indicates that the handle was carved originally with some degree of a back angle—thus it is the earliest example in the study to show other than a straight handle shape. This may only be an example of bias in the sample due to differential preservation.

In 1745, Bishop Spangenberg of the United Bretheren reported from a council meeting of the Iroquois at Onondaga that ". . . At noon, two men entered, bearing a large kettle . . . upon a pole across their shoulders . . . A large wooden ladle, as broad and deep as a common bowl, <u>hung with a hook</u> to the side of the kettle . . ." (Loskiel 1794:138). A third Huntoon ladle (102) has a broken top.

Honeoye Site

The 1750–1779 Honeoye Site (Hne 28-4) yielded a ladle (103) featuring a human face surmounted on back and top by a bear head. (See Plate 15.) On this ladle, it is clear that the original position of the handle with respect to the bowl is at an angle of about 140°. A second Honeoye Site ladle (104) has a broken top.

Townley-Read Site

The sequence of eastern Seneca villages is not clear in the eighteenth century. It is believed that by 1710 the population of the eastern Boughton Hill Site (Can 2-2), burned as a result of the Denonville raid, had moved to the Read Farm Site now owned by Townley (Plp 16-4) (Charles Wray, personal communication). An antler ladle (99) from the Townley-Read Site with a small ring projecting from each side has been preserved. The handle is arched backwards in a curve; the top is broken.

Kendaia Site

An eastern splinter group moved to the Kendaia Site (Ovd 3-3) in Romulus Township, Seneca County, around 1750 (Charles Wray, personal communication). From this site a wooden ladle (105) was recovered with an effigy of a swan or goose with lowered head, as if feeding. The effigy is carved at the end of a handle which is angled sharply backwards. The effigy placement is somewhat similar to effigies on projecting platforms on Eastern Woodlands smoking pipes (J. King 1977. Pictured by Neander, 1624).

Fall Brook Site

The remaining eighteenth century Seneca ladle-producing sites do not fall into a sequence. The Fall Brook Site, also known as Lower Fall Brook, (Cda 4-4), c.A.D. 1750–1775, east of Cuylerville, produced a ladle (107) with the same sharply back-angled handle seen at the contemporaneous Kendaia Site, but the effigy is broken.

Fall Brook, in the Wadsworth Genesee Valley complex, yielded also a ladle (106) with a beaver effigy. The handle projects backward in the shape of an inverted letter "L".

OTHER SITES

The year 1779 marks the date of an expedition sent by the American colonial government to destroy the crops, orchards and villages of the Seneca in western New York. The Seneca had become the allies of the British at the time of the American Revolution and continually harassed the American frontier settlements. After the Sullivan-Clinton-Brodhead expedition accomplished its mission of destruction, the Seneca dispersed, some to the British at Fort Niagara, others to villages in the middle Genesee Valley (Hamell 1979a).

Two post-Sullivan village sites have produced wooden ladles—the Big Tree Site (Cda 58-4) and the Canawaugus Site (Cda 2-3), both c.A.D. 1780–1820 middle Genesee Valley sites.

From the Big Tree Site, the study includes a hawk or eagle effigy ladle (108) with straight handle, and a ladle (109) with a pair of otters surmounting a handle with inverted letter "U" back projection. (See Plate 16.)

Canawaugus village, originally on the east side of the Genesee River c.A.D. 1780, was removed to the west side by c.A.D. 1800. A ladle (110) preserved from this site is in the seventeenth century tradition of carving in one plane and is topped by a small owl effigy. An animal effigy (111) which is possibly a bear and possibly a ladle handle is also from the Canawaugus Site.

Ethnological Ladle

A Seneca ladle (136), dated 1791, collected from Cattaraugus Reservation (MAI-HF) is ornamented with a perching bird effigy which tops a handle with angled back projection. This is the earliest example encountered in the study of a ladle with a bird effigy which is not species-specific.

SUMMARY OF ATTRIBUTES

In summary, of the 22 ladles from the eighteenth century Seneca archaeological sites, there are:

> 1 human
> 1 reclining human figure
> 1 human and bear
> 2 bear or wolf
> 1 pair of otters
> 1 beaver
> 1 hawk or eagle
> 1 owl
> 1 swan or goose
> 1 turtle
> 3 abstract
> 8 top broken

The single eighteenth century Seneca ethnological ladle has one perching bird.

The surviving examples of ladles with back projection handles come from the Huntoon Site, the Kendaia Site, the Fall Brook Site and the Big Tree Site. After mid-century, most handles have rear projections, and most effigies are carved in the round.

The ratio of bowl width to bowl length in eighteenth century Seneca ladles now averages .82, indicating a bowl somewhat wider than long, as compared with the seventeenth century average of about 1.0.

OTHER IROQUOIS LADLES

A few archaeologically recovered eighteenth century ladles from other than Seneca Iroquois sites are included in the study. Among these are two Onondaga ladles from the c.A.D. 1740 Coye Site which show rear projections of the handles. One (113) is topped by a long-legged, long-necked bird effigy (it may have been long-billed), and the other (128) by an animal head. A third Coye Site ladle (131) fragment has a human head effigy at the end of a straight handle. A possible beaver effigy bowl (129) from the Onondaga Sevier Site is dated c.A.D. 1720. (The Coye and Sevier Sites artifacts are from a private collection.)

An ethnologically preserved ladle attributed to the Mohawk and believed to date 1754 is in the collection of Bacone College, Muskogee, Oklahoma. The ladle (130) is carved with an effigy of a seated human holding a ball. The legs and arms are bent so that they touch at knees and elbows. On the head is the outline of a band or turban. The handle projects to the rear in the shape of an inverted "U".

NON-IROQUOIS ARCHAEOLOGICAL LADLES

Archaeological ladles from non-Iroquois sites of the eighteenth century include one (135) of antler, attributed to the Ottawa Algonquian Gros Cap Site, c.A.D. 1710–1760. The effigy is of a pair of facing birds, beaks touching, similar in design to the effigy on the bone comb from the Late Archaic Frontenac Island Site.

Two archaeologically recovered, possibly Chippewa Algonquian ladles from the Black Duck Culture at the Rainy Lake Site (Ontario/Minnesota) could be eighteenth century. One (123) is non-effigy. The other (124) has a carving of two birds perched side by side on the top of a handle with a triangular perforation. The birds have minimal tails (ROM Collection).

NON-IROQUOIS ETHNOLOGICAL LADLES

An ethnological Delaware Algonquian ladle (127), said to date from the 1775 Ticonderoga campaign, is non-effigy; the end of the handle is curved backward. (MAI-HF Collection).

An ethnological Mahican Algonquian "feast" ladle (125) carved with a bear climbing up the handle is attributed to the eighteenth century (NYSM Collection) (Brasser in Trigger 1978:199). The effigy on the ladle is similar in concept to the bear effigy from the seventeenth century Power House Site. The latter, from the evidence of the fracture marks on the feet, also could have been climbing up the handle of a ladle. Another pre-1800 Mahican ladle (126) is non-effigy (MAI-HF Collection).

An ethnological, possibly Potawatomi Algonquian ladle (122), which may be late eighteenth century, features an effigy of an outstretched right hand, palm open and fingers pointing up, tipped with a pointed, triangular shape—perhaps a representation of lightning (DIA Collection).

A Wyandot Iroquoian, ethnologically preserved wooden ladle (121) from the Upper Sandusky region of Ohio dates at least to 1799 and may be older. The effigy is of a seated human, leaning forward, hands on knees. The bowl of the ladle is of the truncated, side delivery type, said to have been carried in the belt by travellers (MAI-HF Collection).

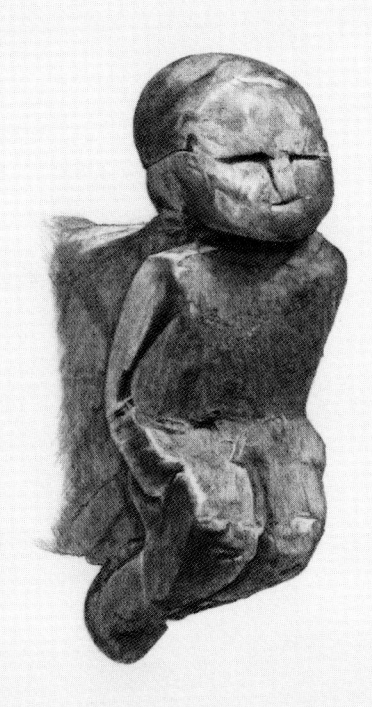

Plate 13. Ladle effigy, wood. *Archaeological.* (91) Snyder-McClure Site (Plp 6-3). Seneca Iroquois, c.A.D. 1687-1710. RMSC AR 18403. About three times actal size. Drawing by Gene Mackay.

Plate 14. Ladle, horn. *Archaeological.* (94) Snyder-McClure Site (Plp 6-3). Seneca Iroquois, c.A.D. 1687-1710. RMSC AR 18589. About one and one-half times actual size. Drawing by Gene Mackay.

Plate 15. Ladle, wood. *Archaeological.* (103) Honeoye-Morrow Site (Hne 33-4). Seneca Iroquois, c.A.D. 1779. RMSC 117/118. About one and one-half times actual size. Drawing by Gene Mackay.

Plate 16. Ladle effigy, wood. *Archaeological.* (109) Big Tree Site. Seneca Iroquois, c.A.D. 1800. RMSC 51/178. About three times actual size. Drawing by Gene Mackay.

CHAPTER III

EFFIGY SYMBOLISM

In the following discussion an attempt is made to understand the effigies on the seventeenth and eighteenth century Iroquois archaeological and/or ethnological ladles in terms of the beliefs and practices of the Native Americans and to suggest possible examples of diffusion or influence from outside sources.

Effigy carving in the Northeastern Woodlands is known as far back as the Archaic Stage of 5,000 years ago and continues into the historic present. Among the Iroquois, effigy carvings adorn ladles, smoking pipes and combs, and also stand alone as figurines and maskettes (Wray 1973; Wray & Schoff 1953; Mathews 1978). The effigy is the most symbolically meaningful physical attribute of the ladles.

It has been observed that among the Northeastern Woodlands Indians, effigies were carved for many reasons. Shimony (1961:161) remarked that among the Six Nations Iroquois, effigies were carved in some instances to serve as identification of the owner. Others were carved by the Iroquois as clan symbols (Speck 1945:84). Some could be interpreted as images relating to one of the secret medicine societies of the Seneca Iroquois (Parker 1909), or for magical or religious uses as among the Chippewa Algonquians (Ritzenthaler in Trigger 1978:749).

Waugh remarks that, among the Iroquois:

> handles of spoons are frequently carved with designs which are ornamental, totemistic, or in response to dreams, particularly those occurring during some indisposition or illness. The dreams are interpreted by a local seer or medical practitioner, who decides upon the design, also the kind of wood, the presentation of such dream-objects to the patient being necessitated to secure recovery. Failure in this respect is believed to be followed by continued illness and eventually by death. The custom seems to have been based upon the belief that the soul can depart from the body and that satisfaction of its desires must be obtained to bring about its return. (Waugh 1916:68).

This echoes the Jesuits' observations among the Huron Iroquoians that the Indians "are convinced that they are affected with diseases only because the soul is in want of something for which it craves; and that it is only necessary to give it what it desires in order to detain it peacefully in the body" (JR 43:267). The Chippewa, too, believe that a representation of the dream-object is necessary in order to secure its protective benefits (Densmore 1929:79-80).

Sagard remarked of the seventeenth century Huron Iroquoians, however, that they carved effigies on their smoking pipes "not for idolatry, but to enjoy looking at them" (Wrong 1939:98), but this would not seem to explore sufficiently the significance of the images made by the New York Iroquoians.

SEVENTEENTH CENTURY HUMAN EFFIGIES

A majority of the surviving New York Seneca Iroquois effigies on seventeenth and eighteenth century archaeological ladles depict humans, either head, torso or figure.

The attention of Europeans to these human effigies (found not only on ladles but also on combs and smoking pipes, and as figurines, maskettes and full size False Face Society masks) focused on their presumed use as "idols." A Christian Mahican Algonquian in 1741 refers to the "idol" of his wife's great grandmother, which was made of leather in the shape of a man and adorned with wampum (Wallace 1951:292). A Dutch journalist in 1634 observed among the Oneida Iroquois the chief's "idol," which was "a head with the teeth sticking out; it was dressed in red cloth" (Wilson 1895:88). The twentieth century practice of covering a mask of the False Face Society when it is not in use may be related to this (Fenton 1941:414). A mask was observed in a Seneca Iroquois village by a member of the 1687 Denonville Expedition (JR 63:289).

Loskiel refers to a Delaware Algonquian "sacrifice" to a head "as large as life," which is "put upon a pole in the middle of the house" (Loskiel 1794:39).

The Quaker missionaires at Cornplanter's Allegany Seneca village in 1798 found the religious activities centered around a post carved in human likeness, decorated with ribbons and trinkets (Deardorff 1956:589).

The Delaware Algonquians reportedly used carved wooden heads as charms and idols (Zeisberger 1910:141). Loskiel called them "manitto" and observed that parents hung these miniatures around the necks of their children as charms, "to preserve them from illness and ensure to them success" (Loskiel 1794:39).

The historical accounts report that these effigies were carved in wood (Waugh 1916:68; Beauchamp 1905:172; Zeisberger 1910:141; Loskiel 1794:39). The archaeological record also gives evidence of human effigies carved in antler, bone, stone and shell (Wray & Cameron MS:n.d.).

Myths and legends of the Seneca Iroquois endow "manikins" with magic power to protect against evil and misfortune (Curtin & Hewitt 1918:219). "Manikins" were also considered to be magic charms used in transformations. (Curtin 1923:232-3).

It appears that the Seneca believed that representations of the human face, figure or limb were endowed with great power, and it seems reasonable to assume that the human images carved on ladle handles carried the same symbolism.

There is evidence for ritual significance in the treatment of head, facial features and hair on the ladle human effigies.

The delineation of a head band on many of the human effigies has been pointed out. Usually the band is depicted as encircling the head, approximately following the hairline, across the forehead and around the nape of the neck. Some of these are incised, others are represented by a ridge in bas-relief.

It is possible that this is a representation of a common headdress. It was observed, indeed, of the seventeenth century Indians of Manhattan Island that "they wear no Hats, but commonly wear about their Heads a Snake's skin, or a belt of their money, or a kind of Ruff made with Deers hair . . ." (Denton 1902:52). Van den Bogaert also noted that "they [the natives of New Netherlands] have long deer's hair which is dyed red, and of which they make rings for the head, . . ." (Jameson 1909:301). Van den Bogaert, however, also observed of the New Netherland Indians that it was the native doctors of the seventeenth century who used snake skins as head gear (Wilson 1895:81-101).

Eliade presents evidence that the minimal requisite for shamanic costume includes "the cap, the belt, the drum . . ." and even among a tribe which did not possess a special shaman's costume, "use is made of a cloth which is wound around the head and without which it would be impossible to shamanize" (Eliade 1964:146-147).

One of the seventeenth century ladles (34) which shows a headband on the human effigy also features a mouth carved in a circular shape with the surrounding area puffed out. The cheeks are sunken. Taken together with the headband, it seems probable that this effigy represents a shaman/doctor in the act of curing or healing by sucking out the disease-causing matter from the patient's body (Mathews 1978:170-171). The Dutch surgeon at Fort Orange is believed to have been the author of a journal of a visit to the Oneida Iroquois in 1635, at which time he observed an Indian doctor engaged in a curing ceremony which involved sucking the patient's neck and back (Jameson 1909:153). Alternatively, the configuration of the mouth on ladle 34 could indicate blowing, perhaps indicating one of the False Faces or beneficent Wind Gods who could blow disease away. Of course, hollowed cheeks might be considered incompatible with this explanation.

A wooden human effigy ladle (59) from the Marsh Site shows the head and torso of a female. The head is warped but appears to show the headband and possibly the protruding lips of the "blowing" curer. The Jesuits recorded an instance of Indian doctors holding hot stones in their teeth and blowing on the patients (JR 14:59-63).

The human head effigy (14) from the Steele Site also has a puffed out area around a narrow, horizontal oval mouth. The effigy fragment does not indicate the configuration of cheeks and head, nor is a headband definable. The effigy is painted or stained light blue. As the color of the Sky Dome (Hewitt 1928:473) or of the Great Bluebird (Hewitt 1928:573), light blue (the third color to be made on earth by the Creator) is a sacred color to many Woodlands people and may indicate a ceremonial context for the ladle.

A headband is delineated on a full figure antler ladle (57) from the Marsh Site, although little other detail has been preserved. The legs and arms are not detailed, although one arm may be seen in profile to the elbow.

Headbands are featured on the human head effigy (113) from the Onondaga Iroquois Jamesville Pen Site and also on one (115) from the Neutral Ontario Grimsby Site.

An antler ladle (13) from the Steele Site has a full figure human effigy which appears to have supernatural attributes. The effigy represents an old (?) female grasping a cane. The effigy is carved in profile and shows a pendulous breast and a notched or serrated spinal column. The head is difficult to interpret, as the features are not clearly executed, and amorphous projections hang from the nape of neck and chin. Fenton notes that in Iroquois ceremonialism, "old men and supernaturals carry canes or staffs" (Fenton 1950:29). Of staffs Fenton wrote:

> The staff is deeply rooted in the Iroquois conception of the ideal older man. 'Old man' has a connotation of affection and respect. Certain classes of supernaturals are called 'Our grandfathers.' This is how the people address, in prayers, both classes of maskers, the wooden faces whom the people impersonate by wearing masks of wood and the Husk-faces; both carry wooden staves . . . In praying . . . the priest says: 'And now your cane receives tobacco, which is a great hickory with its limbs stripped off to the top' (Fenton 1950:30).

The tree which provides the pathway to the upper world is a symbol found often in Iroquois myth and legend. The notched or serrated spine may symbolize the generations of ascent to or descent from an ancestor. It might also represent ascent on the mythical ladder-notched Tree of Life by shamans on trips to the other world-land of the souls (Eliade 1964:487).

Another antler ladle (52), perhaps showing shamanistic attributes, is of a human full figure. Recovered from the Rochester Junction Site, the carving, in profile, shows a notched lower spine. What appears to be the stub of an arm protrudes forward and could have grasped a cane. From the forehead there is a protrusion which may represent a horn.

A wooden ladle (15) from the Steele Site is carved with what may be two stylized human heads positioned back to back. The heads terminate in elongated points, perhaps representing horns.

The two-pronged terminus of the antler ladle (83) from the Boughton Hill Site may represent horns rather than a fork.

The horn of power is a recognized shamanistic symbol (Mathews 1978:172). Among the Iroquois, "horns" were a sign of a chief, or of any being possessing supernatural power.

In summary, the seventeenth century human effigy ladles show characteristics consistent with their interpretation as images of supernaturals. The band around the head, the horn of power, the cane, the exaggerated spinal column and the blowing face, all may be associated with supernaturalism.

POSSIBLE MIDEWIWIN ASSOCIATIONS

The Warren Site cache of two wooden ladles, one (11) with a hawk and the other (12) with a bear, panther or beaver effigy, found nested inside a wooden bowl (10) with a human head effigy, may be evidence of the existence among seventeenth century Iroquois of the Grand Medicine Society or Midewiwin, a life renewal cult of the Great Lakes tribal people. Fenton has recounted the events which led the Iroquois and Huron into contact with Great Lakes Algonquians and Northern Plains Siouans and has considered the possibility that "the Iroquois developed those society rituals, which stress hysteria and possession, during contact with northern peoples . . . the medicine bundle type of shamanistic society which celebrates feasts to propitiate earthbound animal spirits" (Fenton 1941:194).

Skinner's detailed description of the ceremonies associated with the Midewiwin among the Menominee, Iowa, Wahpeton Dakota, Ponca, Bungi Ojibwa and Potawatomi tribes of the Great Lakes region showed the importance of the concepts of throwing, shooting and pointing in the conduct of the ceremonies (Skinner 1920a). It is thus of interest to find among the Iroquois a Medicine Society whose shamans appear to behave similarly. Chief Joseph Logan, an Onondaga Iroquois, discussed with Fenton the history of the Medicine Society among the Iroquois:

> In the old times, perhaps a thousand years ago, men of each Iroquois nation had . . . songs for contesting magic power . . . While they were dancing they demonstrated (their) powers by 'throwing,' or 'shooting sharp objects,' such as 'horns' . . . or one would sing 'something (like a bear) is running around' (Fenton 1942:21).

The song which begins the Round Dance or Medicine Dance is called *gahii'dohon'*, meaning "sharp point, . . . an obscure archaic word" which is used to refer to "sharp objects which shamans shoot" (Fenton 1942:24, note 6).

The effigies from the Warren Site cache may pertain to the "eat all" feast which is a feature of the ceremonies associated with the Mide cult. Society members were said to have owned specially carved wooden bowls and spoons which they brought to the feast. Among the Wahpeton Dakota Siouans, the bowls and spoons were "furnished with handles carved to represent medicine birds or animals" (Skinner 1920a:265). The hawk ladle may represent the symbol for second degree Mide membership as among the Chippewa Algonquians (Densmore 1929:76). The relationship of a bear effigy to the Mide Society would be with the "Bear Below," who was invoked by Skinner's Menominee Algonquian informant: "Be sure and offer tobacco to the Bear Below, reminding him that he owns this medicine . . . for reviving and reinstating the sick" (Skinner 1920a:136).

If, however, a panther effigy rather than a bear effigy was intended (there is indication of a long tail on the animal, although the blunt snout is bear-like), the panther would exemplify an association with the cult of the mystic long-tailed animals. Recently, Brasser (1977:132) has suggested that the Grand Medicine Society had its roots in an Iroquoian dragon (exemplifying long-tailed animals) cult.

The association of the human head effigy bowl with the Midewiwin may be with the spirit of Eeyah, a Siouan god of gluttony (Dorsey 1894:471). Mide bowls were sometimes carved with this human head effigy of Eeyah, a supernatural also described as "associated with the qualities of potency, licentiousness, libertinism and gluttony" (Brasser 1977:132).

If the Warren Site human head effigy is intended to represent Eeyah, it is unusual in that it is not facing the diner, but rather, faces sideways. The Eeyah heads customarily face forward.

The back of this human head effigy is amorphous; the impression is of long, bushy hair being blown back by the wind, introducing the possibility that the Great Flying Head, one of the Whirlwinds, *Dagwanoenyent,* is being depicted (Curtin & Hewitt 1918:187, 261, 474, 481, 485).

Examples of human and animal effigy bowls and spoons associated with the Grand Medicine Society are illustrated in Skinner 1920a:Plate XXII; Dockstader 1961:fig. 221; Dockstader 1973:243; Maurer 1977:fig. 123; Callender in Trigger 1978:650.

EIGHTEENTH CENTURY HUMAN EFFIGY LADLES

Among the eighteenth century Seneca Iroquois archaeological ladles with human effigies, the Huntoon Site produced a small wood carving (100) which may have been a ladle effigy, although it is more probably a maskette. It reproduces an example of a False Face Society mask with downturned mouth, wrinkled forehead and staring eyes with traces of brass inlays. Three tiny holes along one side presumably imitated the attachment holes of the full size masks. The effigy is covered with red pigment. The earliest documentation of masks among the Seneca Iroquois is in 1687 at the Boughton Hill Site (JR 63:289). The Huntoon Site dates 1710-1730.

A ladle (103) from the 1750-1779 Honeoye Site shows an effigy of a human face surmounted by a bear head which may illustrate a curing or masking ceremony. Champlain observed in 1616 elderly Huron Iroquoian women dressed in bearskins, apparently assisting an Indian doctor (Grant 1907:324). Sagard reported "doorkeepers" among the Huron who wore bearskins on their heads, their faces covered up except for the eyes (Wrong 1939:117). A native doctor dressed in a bearskin was observed by John Heckewelder in 1775 at a village on the Muskingum River in Ohio (Wallace 1958:127). Shimony (1961:261) records the existence of a Bear medicine society among the Iroquois.

Other eighteenth century Seneca Iroquois archaeological ladles with human effigies have not been preserved adequately to determine whether or not a headband or other indication of supernaturalism existed.

An ethnologically preserved eighteenth century ladle (130) from the Mohawk Iroquois which dates 1754 features an effigy of a seated human holding a ball. The person appears to have a roached headdress, indicative of a warrior rather than a supernatural.

A no-provenience, probably ethnological wooden ladle (450) has an effigy of a human face surmounted by an eel, salamander or otter head, perhaps bearing an association with a long-tailed animal cult. The date of the ladle is unknown. It has a patina, but as Fenton has pointed out, this is an "unsatisfactory criterion" of age (Fenton 1950:68).

HAND AND FIST EFFIGIES

Ladle effigies of human head and figure appear to have had supernatural or shamanistic associations among the seventeenth and eighteenth century Iroquois and their neighbors. The association of the human hand or finger or fist with power also has been documented both for the New World and the Old World and may have been one of the cultural beliefs held by the earliest Native Americans (Vastokas and Vastokas 1973; Eliade 1964).

Among the seventeenth century Seneca Iroquois, the clenched fist and its probable abstract form, the strapped cylinder, are found as ladle effigies at four archaeological sites, the Steele, Marsh, Boughton Hill and Dann Sites. From the Steele Site, one ladle (16) shows a left hand with four fingers folded into the palm and thumb up. A second Steele Site ladle (17) is stylized, with only the outline of the fist depicted. A fragmentary Marsh Site ladle (60) appears to have had a clenched fist effigy. Also from the Marsh Site is a strapped cylinder effigy ladle (61). The Dann Site ladles include one (37) with five loops encircling a cylinder, another (35) with seven "fingers" enclosing a cylinder, and a third (36) with a solid "strap" wrapped around a cylinder. The Boughton Hill Site example (76) has nine rings encircling a cylinder.

There is some evidence that the Iroquois used the hand or possibly the fist as a metaphor for the League or Confederacy; the clenched-fist ladles could represent this motif. Research on Dutch manuscripts has uncovered a reference to a meeting in 1626 between a messenger of Pieter Minuit, governor of New Netherlands, and the Iroquois. The manuscript reads: one of the sachems "showed me his hand (defective) that his tribe was five times (defective) and yet was one (defective)" (after Charles Gehring, personal communication June 12, 1979).

Parker (1916:11) notes the use of the same metaphor in the Iroquois Constitution, from the fifteenth string of the Tree of the Long Leaves section: "Five arrows shall be bound together very strongly and each arrow shall represent one nation. As the five arrows are strongly bound, this shall symbolize the union of the nations . . ." Parker comments also that King Hendrick in 1755 in a talk with Sir William Johnson used the same allusion.

The clenched fist or ringed cylinder effigies may be derived from a ritual of throwing two sticks to the spectators at the end of a funeral. As described by Sagard,

> Then from the top of the bier are thrown two round sticks, each a foot long and not quite as thick as one's arm, one on one side for the young men, the other on the other for the girls (I have not seen this ceremony of throwing the two sticks performed at all the funerals, only at some), and they go after them like lions to see who will get them and be able to lift them up in the air in their hands, in order to win a certain prize (Wrong 1939:207–208).

This ceremony, again associated with a funeral, is also noted in the Jesuit Relations:

> The Chief puts into the hand of some one of the latter youth a stick about a foot long, offering a prize to the one who will take it away from him. They throw themselves upon him in a body, with might and main, and remain sometimes a whole hour struggling. This over, each one returns quietly to his Cabin (JR 10:271).

A ritualized tug-of-war as a conventionalized element in mortuary ceremonies is documented by Ventur (1980). The wrestling-for-an-object, which is found in the context of mortuary rites among Iroquoians and Algonquians of the Northeastern Woodlands, probably is related to the tug-of-war. The tug-of-war is seen as a transfer of merit from the deceased to the living and, by extension, as a celebration of the continuity of life, despite death. Thus, the Jesuits' report is relevant that at the Huron Feast of the Dead, or Reburial of the Souls, during the seven or eight days spent in preparation, " . . . from morning until night the living were continually making presents to the youth, in consideration of the dead," while the young people were shooting at sticks (JR 10:289).

The Onondaga myth of the Earth Grasper as recorded by Hewitt translates the name of the primordial Good Twin as "He Grasps the Sky with Both Hands" (Hewitt 1928:486). The Creator Twin who is also known as Sapling and thus is the Tree of Life personified, explains, "I myself will continue to grasp with both hands the place whence I came. So that when the time will come when I shall depart from this place I shall just go back to the place whence we started" (Hewitt 1928:486).

The ceremony of grappling for a stick at a funeral may be associated with the concept of grasping the place whence one came at birth and to which one returns at death. The clenched fist or ringed grip would be illustrative of this image.

A depiction of a hand is used as a mnemonic pictograph for the name of the forty-eighth Federal Iroquois chief which translates "he grasps it" (Fenton 1950:67) or "he holds on to it" (Beauchamp 1907:392). This sachemship is held by a Seneca of the Bear Clan, so it is possible that the pictograph may represent a bear paw rather than a human hand.

Elsewhere in the New World, from the Peruvian Moche culture (c.A.D. 400–600), are known pottery vessels shaped in the form of a clenched fist with the knuckle of the second finger slightly raised. One pot of this type, on exhibit in the British Museum, London, is captioned:

> Pottery vessels in this form are related to mountain scenes in Moche art. In some examples the fingers are shown as mountain peaks on one side of the vessel and fingers on the reverse. These vessels are thought to be connected with shamanistic practices and may represent a symbolic gesture used in certain rituals.

Also in the British Museum collection from Peru is a carved bone human forearm and left hand in a clenched fist position with knuckle of second finger slightly raised. A third example is a ceramic fragment showing the elbow, forearm and right hand. The latter two are illustrated in *The Art of Ancient Peru* (Ubbelohde-Doering 1954:fig. 190).

The caption reads:

> The hand is clenched in a fist with the middle finger projecting: position characteristic of the sling, as used by the god of lightning, lord of thunder and rain in the mountains, according to the ancient myths ... It is reported that a statue of Viracocha, a deity of the mountain folk . . . had one fist raised. Viracocha appears as the god of thunder and lightning who smashes his sister's rain-urn so that it rains, snows, or hails. It would be tempting to identify the present fragment as the Fist of Viracocha. However, it is unknown to us whether or not Viracocha, though at one time worshipped throughout the mountains, was venerated also by the people of the Mochica culture; even so there is little doubt that the meaning of the fragment is basically the same as in the case of Viracocha's fist.

The clenched fist frequently is seen in nineteenth century Euroamerican scrimshaw carvings, made primarily from the teeth of deep sea sperm whales. Sperm whale fishing did not begin outside of American waters until the eighteenth century (Flayderman 1972:17), by which time surviving archaeological examples of wooden ladles show that the Seneca had been carving clenched fists for at least 100 years.

There is evidence that the Seneca participated in nineteenth century deep sea whaling. The son of a nineteenth century Cattaraugus Seneca whaling crewman preserved his father's collection of scrimshaw (Dodge 1951:4). A journal written by a nineteenth century Seneca Iroquois man while aboard a whaling ship is on loan to the Rochester Museum collections.

The clenched fist symbol is known also from European heraldry (Grant 1924:66; Clark 1892:plate XL, XIII), but its earliest appearance is not known.

It is possible that the clenched fist represents a gesture of defiance. Such an interpretation would not be out of place for the seventeenth century Seneca.

RECLINING HUMAN FIGURE EFFIGIES

A different use of the human figure as an effigy is seen in the reclining figure ladles. These carvings show a human figure lying on its back, legs slightly bent at the knees, arms at the side, head tilted back.

The earliest example among the Seneca is found on an antler comb (RMSC 1388/24) from the 1645–1660 Power House Site. There is a fragmentary example on a wooden ladle (62) from the Marsh Site, a fractured antler example (38) from the Dann Site, a wooden specimen (75) from the Boughton Hill Site and another wooden one (91) from the Snyder-McClure Site.

It is possible that this effigy represents a sacrificial motif, similar in concept to the prehistoric Central American "Chac-Mool" statues carved by the Toltecs c.A.D. 900–1200, which show a human figure in a reclining position. The Chac-Mool represents the divine messenger who received and carried to the Sun the offering of human sacrifice.

A motif possibly related to this is seen in a human effigy ceramic bowl from the Temple Mound I culture which dates c.A.D. 700–1200 in the Eastern Woodlands (Willey 1966:250). In this bowl (RMSC AR 17669), the body of the human figure forms the concavity of the bowl, the head protrudes from one side and the legs and feet from the other, and the arms are held at the sides. In both the Chac-Mool and Temple Mound I examples, the head is alert as in a living person, whereas the ladle effigies show the head tipped back either as if in a trance or dead.

Fenton suggests that the pictographs on the mnemonic cane used in the roll call of the Iroquois chiefs and which show a prostrate figure "illustrate that part of the introduction to the Eulogy which sings of the founding chiefs lying in their graves on the laws that they legislated" (Fenton 1950:29). It is possible that the reclining figure effigies are three-dimensional illustrations of this theme.

This explanation would see the reclining figure effigy as a symbolic portrayal of the "attitude of death," that is, with the face upward (Beauchamp 1922:37).

It may be only coincidence, but the earliest preserved examples of ladle and comb reclining figure effigies date to the earliest Jesuit/Seneca contacts. There is a striking resemblance of the effigy to the dead Christ figure in Michelangelo's *Pieta*. A major difference is that in the *Pieta*, the right arm of the Christ figure hangs straight down, while in the ladle and comb effigies the arms are held at the sides. The orientation of the figure is, however, the same in the *Pieta* and the ladle effigies, with the head hanging to the left.

In the Rochester Museum collections, Jesuit rings with the *Pieta* theme have been preserved from the Power House, Marsh, Boughton Hill, Rochester Junction, Snyder McClure, Kirkwood and Huntoon Sites (Wood 1974). Of the sites which produced reclining figure effigy ladles, only the Dann Site has not produced a Pieta ring or medal. Other Jesuit material, however, is present at the Dann Site, and the lack of Pieta rings is likely an accident of differential collection.

The reclining figure effigies, including the one on the comb, cover a time span between c.A.D. 1645–1660 and c.A.D. 1687–1710, a period not too long for one carver to have produced all of the examples. The reclining figure effigies, however, are found both in eastern and in western Seneca villages. While this might indicate a change of residence by the carver, it is more probable that inter-village trade is responsible. That the images are very like each other might indicate that multiple carvers each copied directly from a Pieta ring rather than that all the effigies were carved by one person.

The reclining figure effigies thus may exemplify multiple symbolism in which traditional images are assimilated to European motifs with which they are compatible. Analogously, the acceptance by the Seneca of the Christian concept of Christ as the resurrector may have been facilitated by the Iroquois tradition of the primordial Good Twin, Sapling, as resurrector, and the belief in death as rebirth which was shared both by Christian Europeans and non-Christian Iroquoians. (Hamell personal communication).

HUMAN BUTTOCKS EFFIGY

Another example of the use of a portion of the human anatomy as an effigy is illustrated by a ladle (101) which may represent human buttocks and which was recovered from the early eighteenth century Huntoon Site. It could refer to the "buttocks watcher" (Thompson 1966:296), a theme familiar to the native Woodlands people. In these legends, "a trickster commands his buttocks to act as watchman while he sleeps. The tale usually represents the buttocks as talking" (Thompson 1966: 296, note 83).

HOURGLASS EFFIGIES

The hourglass symbol, seen on several of the ladles and combs, can be understood as an abstraction of the human figure. The hourglass design, which has worldwide distribution, represents a stylized, headless human figure and may be used in a genealogical pattern. When the hourglass figures are linked horizontally, they depict affinity; when they are linked vertically, they depict descent. George Hamell (1979b) has discussed the significance of the hourglass design in Iroquois wampum belts, referencing the Carl Schuster manuscripts being edited by Edmund S. Carpenter.

In the panther effigy ladle (68) from the Marsh Site, the animal is depicted with legs bent at the knees in an anatomically impossible position, but the effect is that the space between the front and back legs of the animal becomes an hourglass shape. A bear effigy comb from the Adams Site (c.A.D. 1550–1575) also has the legs angled so that the hourglass shape is seen in the open space.

The Marsh Site produced a small ladle (70) on which the hourglass shape itself is carved as the finial. A similar example in antler (94) from the Snyder-McClure Site is more elaborate in that it is rayed; that is, there are protruding spikes in addition to the basic hourglass shape. A Cayuga Iroquois antler comb, c.A.D. 1670–1710, has an engraved rayed hourglass in addition to a carving which incorporates an arbor, serpents, and hockers, probably representing "an important mythological theme, perhaps "'origins'-related" (Hamell 1979b:fig. 5b).

In the 1754 Mohawk Iroquois ladle (130) showing a seated human figure effigy, the legs and arms are bent and touch at knees and elbows so as to outline an hourglass shape.

An hourglass design also may have been intended on an antler ladle (134) from the prehistoric Cayuga Iroquois Genoa Fort Site.

BEAR OR WOLF EFFIGIES

Ladles featuring a bear or wolf effigy are second in frequency of appearance on ladles in the study. It is not always possible to distinguish between the effigies of the two animals, except when a tail is fully represented.

Bear and wolf are familiar clan animals among the Iroquois as well as other Woodlands tribes (Beauchamp 1916:69; Curtin & Hewitt 1918:44; Coleman 1925:21; Fenton 1951:49). The forty-eighth chief in the Federal list of the Iroquois League is a Seneca whose name translates "he grasps it" (Fenton 1950:67). The illustrative pictograph is sketchy, but appears to represent either an open hand with fingers extended or a bear paw and claws (also, conceivably, a turtle foot). As the Bear Clan is the owner of this Sachemship, it is perhaps more probable that a bear paw is illustrated.

Bears also figure prominently in ceremonialism among many of the Woodlands people. Bear ceremonialism also seems to have been documented for the prehistoric period (Ritchie 1947; 1950; Hallowell 1926).

Curing is associated with bears. As observed by Loskiel, among the Delaware and Iroquois, a doctor dressed in bearskin performed "with calabash in hand, singing and dancing and scattering hot ashes, sleight of hand tricks with pieces of wood and making horrid noise" (Loskiel 1794:111). In 1775, at a large Indian village on the Muskingum River, Ohio, Heckewelder observed a native whose dress "consisted of an entire garment or outside covering, made of one or more bear skins" (Wallace 1958:127). Heckewelder was told that the man was a doctor. At a curing ceremony among the Huron Iroquois witnessed by the Jesuits, a man put a large live coal in his mouth and growled like a bear into the ears of his patients (JR 14:59-63). Champlain reported in 1615 a cure which could be effected by summoning "large number of men, women, and girls, including three or four old women. These enter the cabin of the sick, dancing, each one having on his head the skin of a bear or some other wild beast, that of the bear being the most common" (Grant 1907:324). Persons who become possessed as a result of watching a dance of a certain medicine society, for example, the Bear Society, will then have to be cured by that society and will themselves become members (Fenton 1940:421).

It is possible that some of the bear effigy ladles observed in this study may be evidence of their owner's membership in such a curing group or possibly the individual's role as a shaman/doctor.

Bear or wolf effigy ladles have been preserved from the Factory Hollow, Warren, Marsh, Power House, Boughton Hill, Rochester Junction and Canawaugus Sites.

The Power House Site example (29) appears to have been broken off some other object as evidenced by the fractured undersides of the feet. Around the body, slight indentations appear to be marks made by string or cord. The significance of these marks is not known. A Mahican "feast" ladle (125) (New York State Museum Collection) has an effigy of an animal climbing up the back of the handle (illustrated in Trigger 1978:199, fig. 3). Willoughby figures an animal with a bear-like snout climbing up a pipe (Willoughby 1935:182, fig. 104f). The Power House Site bear effigy could have been positioned similarly on a ladle.

The antler bear effigy ladle (54) from the Rochester Junction Site shows the bear sitting upon its haunches with its paws up, perhaps illustrative of the Iroquoian belief in a monster bear whose "life is assailable only in the soles of his feet . . . as he raises his foot . . . you will see a white spot in the sole; there is his heart" (Curtin & Hewitt 1918:258-9; 799-800 n. 153). The bear is carved in profile, in outline only. It is similar to the bear effigy figured by Quimby (1966:126, fig. 27B) which is from the 1710-1760 Ottawa Algonquian Gros Cap Site. The Ottawa effigy is of a standing bear with an arm outthrust and holding a cane, perhaps also related in imagery and concept to the 'old woman with cane' effigy ladle (13) from the Steele Site. It will be recalled that Fenton (1950:30) associates canes with supernaturals in Iroquois beliefs.

OTHER MAMMALIAN EFFIGIES

Other mammals carved on seventeenth and eighteenth century Seneca Iroquois ladles include a panther (68) from the Marsh Site, an otter (30) from the Power House Site and a pair of otters (109) from the Big Tree site. A weasel effigy ladle (118) was preserved from the Ontario Neutral Iroquoian Grimsby Site (pre-1650). These animals may be allied to a class of mystic, long-tailed animals which are not fully understood. Brasser suggests "cultic associations" for "effigy pipes showing snakes, lizards, salamanders, and other longtailed animals. All these animals were representations of a breed of giant dragons, called *Ontarraoura* (Huron), believed to live underground and in deep water" (Brasser 1977:130). Curtin & Hewitt (1918:797-798, n. 135) remark on "one of the most firmly held beliefs of the Seneca and other Iroquoian peoples" in a giant horned serpent who lived in deep rivers and lakes. Hamell (1980b) has explored the importance to Iroquoians of the "Rattlesnake Man-Being as Creator."

Among other animals carved as ladle effigies in the seventeenth and eighteenth century Seneca group, the beaver is represented on two ladles (39, 40) from the Dann Site and one (106) from the Lower Fall Brook Site. One of the Dann Site ladles (39) shows a swimming beaver (otter?) effigy

(New York State Museum Collection). An antler comb from the Boughton Hill Site also shows a swimming beaver or otter. (Also see swimming beaver/otter pictograph engraved on bark Midewiwin mnemonic scroll (Hoffman 1891:292; Mallery 1893:241). It is possible that the beaver also may be associated with the long-tailed animal cult.

A horse head effigy appears on a wooden ladle (19) from the Seneca Iroquois Steele Site. An Onondaga Iroquois ladle (128) from the Coye Site also has a horse head effigy (private collection).

REPTILIAN EFFIGIES

Eight turtle effigies have been preserved in the seventeenth and eighteenth century Seneca Iroquois sample. Most of these are fragmentary. Seven (20, 21, 69, 80, 81) are of wood. Two (41, 93), which are brass inlaid, more probably are wooden smoking pipe effigy fragments. One turtle effigy (82) ladle is of shell. The importance of the turtle in Iroquois life is attested by its central position in the Creation Myth (Hewitt 1903) and its long history as a clan symbol. Turtle effigies were recovered from the Steele, Marsh, Dann, Boughton Hill and Snyder-McClure Sites.

A ladle effigy (22) from the Seneca Iroquois Steele Site is probably of a snake or eel. A wavy-edged handle of a ladle (116) from the Neutral Iroquoian St. David's Site, Ontario, may have represented a snake or eel. These ladles could have had clan or cult association.

BIRD EFFIGIES

Our knowledge of aboriginal use of birds as decorations in prehistory and in the seventeenth and eighteenth centuries is derived from archaeology, reports of visiting travellers and missionaries, and the myths and legends of the native people.

The bird effigy is known archaeologically as far back as the Late Archaic Stage in New York State. An antler comb (RMSC AR 36032) carved with an effigy of two facing birds, beaks and breasts touching, was recovered from the Frontenac Island Site (c. 4930 ± 200 B.P.). The protohistoric Seneca Iroquois Reed Fort Site produced an antler snipe effigy comb (RMSC AR 19223). An antler comb (private collection) attributed to the historic Cayuga Iroquois Genoa Fort Site continues the tradition, with an effigy of two facing birds, beaks touching, possibly a Snipe Clan illustration.

Among the historic Seneca Iroquois, archaeologically preserved hawk or eagle ladle effigies are known from the seventeenth century Warren (11), Marsh (64) and Rochester Junction Sites (53) and from the eighteenth century Big Tree Site (108). An archaeologically preserved owl effigy ladle (51) was recovered from the seventeenth century Markham Site and another (110) from the eighteenth century Canawaugus Site. An archaeologically preserved long-necked, long-legged, long-billed heron, crane or bittern antler ladle (63) was recovered from the seventeenth century Marsh Site. A long-necked swan or goose effigy ladle (105) was recovered from the eighteenth century Kendaia Site.

A long-legged water bird effigy (133) also is known from the eighteenth century Onondaga Iroquois Coye Site. An antler ladle effigy (135) of two facing birds with bills touching was recovered from the eighteenth century Ottawa Algonquian Gros Cap Site. A wooden ladle (124) from the eighteenth century Chippewa Algonquian Rainy Lake Site shows two perching birds, side by side.

One ethnologically preserved ladle (136), dated 1791, was collected at the Cattaraugus Reservation and is presumed to be Seneca Iroquois. The effigy is of a perching bird.

The Jesuits understood the seventeenth century Huron Iroquois to believe that they had two souls. One remained in the bones, while the other separated itself from the body at death, remaining nearby until after the Feast of the Dead at which time it turned into a turtle dove and went to the land of the souls (JR 10:287).

A visiting Dutch journalist observed in 1634 a large wooden bird which was carved on top of the palisade at the entrance to the grave of an Oneida Iroquois Chief (Wilson 1895:92).

James Adair reported that the eighteenth century southeastern Indians did not use birds "as augurs of the future. They esteem pigeons only as . . . food, and they kill the turtle-dove" (Adair 1775:26). They also had "birds," however, carved on top of their temples (Swanton 1946:613).

The Iroquois, too, used the migratory passenger pigeon as food. Writing of the captivity of Sarah Whitmore by the Indians in 1782, Mrs. Sarah Gunn reports:

> about this time occurred the assembly of all the tribes at what was known as the 'Pigeon Roost.' Near the shores of Seneca Lake was the rendezvous of thousands upon thousands of pigeons at mating and nesting time. For this reason, annually, the Indians assembled here for days and weeks together. The young birds were fat and juicy, and were devoured in large numbers; while the squaws smoked and cured great quantities of these for future use. Consequently, with the Indians, the 'Pigeon Roost' was synonymous of a feast and a dance, and especially of a council (Gunn 1782:517-518).

Bird bones have been found in pots and kettles associated with Seneca Iroquois human burials, probably as food for the deceased. A skeleton of a great blue heron was found in a prehistoric Iroquois human burial on the Markham and Puffer estate near Avon, but the evidence suggests ceremonial interment rather than food. The burial may have been of a matron of the Heron Clan. Although some clan birds were eaten, it is believed that the heron was not (Wray 1964:11-12).

Certain birds play central roles in Iroquoian myth and legend. The mystic Dew Eagle, for example, in its role as the chief of the Sky World, maintains on its back a pool of dew water. It was with this magical water that the bird glued on the severed scalp of the legendary Good Hunter, thus restoring his life and giving to mankind the boon of the life-restoring medicine preserved by the Little Water Medicine Society (Beauchamp 1901:154-155).

Owls are characterized as witches or wizards in virtually every story in which they appear (Curtin & Hewitt 1918:116, 126, 134, 205, 212). Seventeenth century northern Iroquoians ascribed the origin of curing feasts to a meeting of wolves and the owl at which time the owl predicted the coming of the underwater dragon (JR 10:325, n. 17).

In the Iroquois Creation Myth, swan or geese flew up to intercept Sky Woman as she fell from the Sky World toward the primeval water below. The bittern in its characteristic pose with head pointing skyward saw her falling, while the loon, looking down into the water, saw her reflection and concluded she was rising from the depth of the water (Hewitt 1903:179-180).

SUMMARY

It is not possible to know to what extent the existing seventeenth and eighteenth century Seneca ladles accurately reflect the society of which they were a part, as there is an unmeasurable amount of bias due to differential preservation and collection.

The ladles which are extant, however, overwhelmingly portray power symbols and repeatedly use life-over-death or life renewal symbolism. In so far as the documentary sources confirm the symbolic messages of the ladle effigies, it is possible to infer a society in which the sacred and the secular are intertwined inextricably. The power and influence of the shaman/doctor is clearly revealed in the human effigies with shamanistic headbands, horns of power, serrated spine, masking and bear ceremonialism. The mystic animals and the mortuary grappling sticks celebrate life renewal in the face of death, as do the reclining chiefs of the League whose laws will live after them.

The birds and animals serve dual purposes as secular clan symbols and sacred power symbols (the turtle which bears the world on its back; the Dew Eagle which freshens the earth; the doctor/bear, and the bird as soul).

The hourglass serves by symbolically describing the social organization as it appears in the negative space of a power-animal's legs.

CHAPTER IV

NINETEENTH CENTURY ETHNOLOGICAL LADLES

The list of nineteenth century ethnological ladles which follows will show considerable change in the attributes of effigy type, bowl shape and handle angle. The nineteenth century ladles were collected from Indian reservations and are grouped according to reservation, although it is understood that they do not represent homogeneous groups any more than their predecessors in the seventeenth and eighteenth century villages. Because of the great number of ladles presented in this chapter, they are listed in tabular form.

The study includes ethnological ladles from five of the six New York Iroquois tribes, a sample from the eastern Algonquians, a group from the northern Algonquians and a number from the Great Lakes tribes.

The largest group is designated Seneca Iroquois and includes examples from Cattaraugus, Tonawanda and Allegany Reservations in New York State, as well as a sample from the mixed group in Oklahoma which was called Seneca. Smaller samples of Cayuga, Onondaga, Oneida and Mohawk ladles are included. There is also a group marked Iroquois, New York, another marked Iroquois, Canada, and one identified as Iroquois, Grand River.

The Appendix contains a tabulation of the ladles in the study which includes collection name and number.

It is not possible to be certain that the ladles in this group are entirely of nineteenth century manufacture. Some could be eighteenth century or earlier heirlooms. Others which were collected early in the twentieth century could have been new at that time. (A few ladles which were definitely of twentieth century manufacture were excluded from the study.)

EFFIGY MOTIFS

Seneca

Cattaraugus Reservation Seneca

Three animal heads (177, 178, 179).

One bird head with hooked beak (similar to the antler heron/crane/bittern ladle from the Marsh Site) (174).

One pair of animal heads joined back to back. The only double effigy ladle previously seen was the human/bear combination from the Honeoye Site (176).

One bird head (173).

One sleeping, long-necked swan/goose (140). The Kendaia Site example (105) was in feeding posture.

One sleeping bird, head and neck curled back into tight loop (142).

One perching owl, full figure, previously seen only as a head (141).

Two climbing human figures, holding placards in outstretched arms (one placard incised "May 22, 1872") (137, 138).

One seated human figure, legs hanging down, hands on knees. A head band is outlined (139).

One buffalo (180).

One human astride buffalo (175).

Thirty unidentifiable perching birds. The tail of the bird provides the means to hook the ladle to the rim of the pot (143-172).

Tonawanda Reservation Seneca

One animal head (198).

One sleeping bird, head and neck curled back into tight loop (195).

One spotted animal ("carved in 1880 for the county fair") (196).

One cat (197).

One conventionalized perching bird (194).

One seated human eating from a bowl, being watched by seated dog (193).

One seated human pulling on a boot. (Handles of latter two ladles intricately carved into lyre shape with vase in central opening) (192). (See Plate 17.)

Cattaraugus or Tonawanda Reservation Seneca

One fox or wolf head (215).

One perching bird (probably owl) (202).

One perching bird with long beak (probably snipe) (201).

One seated human holding placard (incised "May 22, 1877") (200). (See Plate 18.)

One ball/disc (216).

One abstract motif (probably a perching bird) (203).

Eleven conventionalized perching birds (204-214).

Allegany Reservation Seneca

Six conventionalized perching birds. (One of these was recovered from the Cornplanter Cemetery (Abrams 1965) and could date to the late eighteenth century) (220-225).

Seneca

One human leg and foot (Lewis Henry Morgan Collection—RMSC). Although no other ladles with such an effigy were encountered in the study, the effigy has antecedents believed to date to the seventeenth century Iroquois. The leg and foot effigy was carved on the distal end of three wooden ball-headed clubs. One of these is in the National Museum of Denmark at Copenhagen. A second is in Skokloster Castle Museum, Sweden and a third in the Swedish Ethnographical Museum, Stockholm. Another possible model could be bear molar teeth which were modified by the Iroquois to resemble a human leg and foot. The human leg and foot pictograph is known from heraldry and is also a common scrimshaw motif (231). (See Plate 19.)

One abstract design (271).

One sleeping bird, head and neck curled back into tight loop (234).

Two sleeping swan or goose (232, 233). (See Plate 20.)

Double wolf (?) heads back to back, sharing one pair of ears. (Incised "Feb. 8, 1901"; included with nineteenth century for comparison) (261).

One combination human/bear, marked "PB 1892". The human head faces the front of the ladle, and has an elongated chin. The bear face is on top of the human head, face up (228).

One human figure on hands and knees, with indication of a headband, may illustrate a person possessed by a spirit as observed by Champlain in 1616. These people who walked "on all fours like beasts" were cured by a society of masked men who blew upon them shaking turtle shell rattles and making loud noises (Grant 1907:326) (Biggar 1929:153-155). The patient automatically became a member of the curing society (227).

One ladle effigy of two pairs of acrobats, (Lewis Henry Morgan Collection—RMSC) (229).

One effigy of two wrestlers, now broken (LHM Collection—NYSM) (230).

One squirrel eating a nut (LHM Collection—NYSM). In Iroquois belief, the squirrel is associated by its long tail with the mystic animals (262).

One sprawled frog (?). The frog is familiar in Iroquois legends and also is associated with medicine by the Dakota Sioux. Densmore illustrates a ladle from the Sioux with a frog effigy which "belonged to a medicine man who 'doctored' under the tutelage of the frog" (Densmore 1948:Plate VII) (263).

One ball, may indicate a Ball Clan owner (264).

Twenty six generalized perching birds (235-260). (See Plate 21.)

Oklahoma Seneca

One wolf/bear head (287).

One animal head (289).

One-eared animal head. Ladle handle curves forward in "C" shape (288).

One bear, on pedestal, chip carved in geometric design. Dodge theorized that this technique was learned by the Maine Penobscots from Scandinavian lumberjacks after 1870 (Dodge 1951:2-3). Chip carving is seen earlier, however, on New York Seneca Iroquois artifacts, as on a mid-nineteenth century stirring paddle (LHM RMSC) and a comb from the Canawaugus Site (c.A.D. 1800) (RMSC) (282).

One seated dog or wolf. Chip carved (283).

One stylized bear. Ladle handle curves forward in "C" shape (284).

One otter (285).

One human/turtle combination. Human head has elongated chin. Turtle is on back of the human head (273).

Three bird heads. One has handle curved forward in "C" shape (279, 280, 281).

Three sleeping swan/goose (276, 277, 278).

One buffalo (286).

One seated human, painted and pencilled to indicate breechcloth and leggings. Ribbon around neck as scarf (272).

Two conventionalized perching birds (274, 275).

Other Iroquois

Although there are fewer nineteenth century ladles from the other New York Iroquois tribes included in the study, a summary of these is included to compare with the nineteenth century Seneca Iroquois.

Cayuga

One snake (310).

One turtle (309).

One bird head (308).

One abstract design (possibly perching bird) (307).

Two generalized perching birds (305, 306).

Two with a trio of ducks (302, 303). (See Plate 22.)

One with a pair of ducks (304). The effigies in the latter three ladles are shown sitting side by side. The handles are chip carved with a series of serrated lines on the fronts of the ladles. There is a rectangular perforation of the handle. The sides of the handles are incised with stick figures which may be genealogical abstractions.

Two climbing human figures, one holding a placard, the other with arms outstretched to the edge of the handle. The latter is carved with a spoon-lipped mouth as in a mask of the False Face Society (300, 301).

Onondaga

One turtle (339).

One bear (337).

One perching owl (324).

Two bird heads (330, 331).

Two sleeping swan/geese (322, 323).

One sheep (334).

One rabbit (336).

One buffalo (335).

One abstract arched design (341).

One abstract animal (buffalo?) (338).

One abstract bird head, large eye perforation, square beak (343).

One cage-enclosed ball (340).

One human figure arched in back bend position (320). (See Plate 23.)

One human head and torso, with head arched back, perhaps illustrating ecstatic trance. The effigy is evocative of A.F.C. Wallace's description of the preacher of the Code of Handsome Lake at a 1951 Six Nations Meeting on the Allegany Reservation, "As he (the preacher) speaks, his eyes are half closed and his head is tilted back, and he holds cradled before him in both hands a thick rope of white wampum strings" (Wallace 1972:9) (319).

One human, seated in chair, reading a book. Arms and legs are bent to touch at elbows and knees, outlining an hourglass shape (318).

One reclining human figure, knees bent, holding a jug, perhaps representing the theme found among the Wyandot on pipes which is said to refer to the consumption of large quantities of rum during the annual ceremonies of the Lion or White Panther Society (Brasser 1977:130-131) (321).

Two ladles with animal (dog, monkey, raccoon?) arched over a sleeping swan/goose. One is illustrated in Beauchamp (1905:opp. 238, Plate 21, fig. 102). A duplicate is in the Field Museum of Natural History, Chicago (332, 333).

Five generalized perching birds (325-329).

Oneida

One stylized human face (345).

One human head and torso (344).

One bear (352).

One clenched fist (346).

One sleeping swan/goose (348).

Three long-billed birds (identified as grouse or woodcock) (349, 350, 351).

One abstract design (357).

One squirrel (353).

One perching bird (347).

Mohawk

One beaver (361). (See Plate 24.)

One quadruped (360).

One seated human (366a). (See Plate 25.)

Tuscarora

No effigy ladles identified in this study.

New York Iroquois

One object (human?) with spherical head and spherical body and one pair of arms outlined on the sides of the handle. There are no legs nor are hands indicated. Possibly this is a visualization of the magical stone which occupied a special position within the Midewiwin lodge. As described by Skinner's Menominee informant, "At the east where we first enter there was a small wigwam containing a stone, and this became blue clay (color). The stone that was seated there was a moving power, a hero too, though it had no hands or legs. It was round, but it spoke like a human being" (Skinner 1920:145) (385).

One abstract design, possibly a head with scalloped top (386).

One sleeping swan/goose (368).

One cock's head with cockscomb (369).

Nine generalized perching birds (370–378).

Six Nations Iroquois

One snake (399).

One dog/Wolf (397).

One clenched fist (ladle has side delivery bowl) (387).

Two bird heads (395, 396).

One cat (398).

Seven generalized perching birds (388–394).

Canada Iroquois

One pair of animals (411).

One pair of crouching, spaniel-type dogs (410).

One abstract design (bird head?) (412).

Two sleeping swan/goose (405–406).

One feeding swan/goose (407).

One perching bird (408).

One buffalo (409).

Iroquois

One animal (441).

One bear (436).

One animal head (possibly wolf or dog) (440).

One horse head (439).

One perching owl (423).

One abstract design (443).

Three sleeping swan/goose (432, 433, 434).

One feeding swan/goose (435).

One bird head (431).

One ball/disc. (442).

One rabbit (438).

One buffalo (437).

One human figure, head and back arched (421).

One small human figure shown climbing up the back of the neck of a human head which has an elongated chin, perhaps a representation of the widespread "clinging person" myths (Curtin & Hewitt 1918; Levi Strauss 1978) (422).

Seven generalized perching birds (424–430).

Unknown Provenience but Probably Iroquois

Three bear/wolf head (465, 466, 467).

Two bird head (463, 464).

One human/salamander combination (See Parker 1910: Plate 18—ladle attributed to Delaware) (450). (See Plate 26.)

Twelve perching birds (451–462).

Eastern Algonquian

Delaware

One human face (468).

Two abstract (a wrapped cylinder and a possible snake) (469, 475).

Nanticoke

One sleeping swan/goose (482).

Mohegan

One human face (485).

Northern Algonquian

Abenaki

One turtle (497).

Cree

One ball (508).

One beaver (507).

Great Lakes Tribes

Potawatomi

Two bear (521, 522).

One animal head (523).

One bird head (520).

Five abstract designs (524–528).

One open hand, fingers tipped with triangular projection (may be eighteenth century). The human hand as a power-filled object is the subject of an Iroquois legend in which the hero uses the power of a representation of a tiny human hand to ward off evil (Curtin 1923:4) (122).

Winnebago

Three abstract designs (566, 567, 568).

One eared animal head, described on catalog card as "medicine spoon. Used in mixing and administering love potion" (565).

One skunk (563).

One pig head (564).

Twelve bird heads. Ten of these have forward projection "C" shaped handles. Two of these are carved in the manner of the pop-eyed Early Woodland Stage bird stones attributed to the Meadowood Culture (Ritchie 1965:190). Pop-eyed (551, 552). Other "C" shaped (553-560). Bird head, not "C" shaped (561, 562).

Sauk & Fox

Four abstract designs (604-607).

One bear (594).

Two human (one said to be used in Midewiwin ceremony has flat knob on top of head) (587, 588).

One-eared animal head (596).

Four stylized bird heads (590-593).

Three horse heads (597, 598, 599).

One buffalo (600).

One turtle, said to be used in "war bundle feast" (602).

One animal (601).

One stylized porcupine (?) (595).

One finger (?), carved as bent at two joints. In certain Iroquois legends, magic power resides in a single finger which is used by the mythological Stone Giants to find their enemies (Curtin & Hewitt 1918:111) (589).

Menominee

One abstract design (631).

One stylized bear (629).

One rabbit (630).

Mascouten

One otter (653).

One stylized beaver (652).

One ball-headed club (654).

Ojibway

One bird head (664).

One clenched fist (660).

One long-legged bird (661).

Two perching birds (662, 663).

Chippewa

One bird head (691).

One long-legged bird (689).

One ball-headed club (692).

One perching bird (690).

Ottawa

One animal head (698).

One perching bird (697).

One seated human (696).

Among the eastern and northern Algonquians, although the sample is small and probably biased, there appears to be considerably less effigy carving than among the Iroquois. The Great Lakes tribes are more prone to effigy carving but still carve less than the Iroquois in general and much less than the Seneca Iroquois in particular.

Of 29 ladles in the eastern Algonquian sample, only 5 have effigies. Of 23 ladles in the northern Algonquian sample, 3 have effigies. Among the 179 Great Lakes ladles 84 have effigies.

By contrast, among the Iroquois, 239 ladles of a total of 314 have effigies. The New York Seneca Iroquois are most likely of all to carve effigies: 113 out of a total of 134 ladles have effigies.

TABULATION ACCORDING TO LOCUS OF COLLECTION

A tabulation of four samples of ladles from four different nineteenth century geographical locations points up the extent of Seneca effigy carving. The study sample provided 56 ladles from the Cattaraugus Reservation, New York; 42 ladles from the Six Nations Reserve, Canada; 59 ladles from Oklahoma and 83 ladles from Wisconsin. It is recognized that the Wisconsin group was not taken from as nucleated a location as the others, but was included for comparison because of its high percentage of ladles from Great Lakes tribes. As previously observed, the amount of bias in the sample is not known. There could have been a tendency to prefer to collect effigy ladles rather than non-effigy ladles.

Cattaraugus Reservation, New York

Seneca	53	
Cayuga	2	Of the total,
Unknown	1	92% effigy.

Six Nations Reserve, Canada

Seneca	1	
Cayuga	14	
Tuscarora	1	
Mohawk	2	Almost 100% Iroquois,
Oneida	5	but only 64% effigy.
Iroquois	18	
Nanticoke	1	

Oklahoma

Seneca	23	
Seneca Res.	3	Although 50% Iroquois and still using
Iroquois	3	conventional Iroquois effigy subjects,
Cayuga	2	only 50% effigy.
Delaware	3	Iroquois: 83% effigy
Sauk & Fox	23	Others: 17% effigy
Wyandot	1	
Potawatomi	1	

Wisconsin

Oneida	10	
Stockbridge	2	Innovative subject matter but only
Chippewa	2	34% effigy. While 7 of 10 Oneida are
Mistassini	2	effigy and 12 of 27 Winnebago are
Potawatomi	17	effigy, only 3 of 17 Potawatomi are
Menominee	22	effigy and 3 of 22 Menominee.
Winnebago	27	Others totally non-effigy.
Ojibway	1	

The Winnebago are particularly innovative as compared with the New York Iroquois, with stylized bird heads on "C" shaped ladles. Two of these are carved with the "pop eyes" characteristic of the bird stones in the Meadowood Phase of the Early Woodland Stage of New York State (Ritchie 1965:190).

Seneca in Oklahoma carve bears, humans, buffalo, otter, swan/goose, animal heads and a human/turtle combination. There are only two perching birds plus a bird head "C" shaped ladle. The sample differs from its Cattaraugus Reservation counterpart primarily in the small number of generalized perching bird effigies.

The seventeenth and eighteenth century New York Seneca were known to be a diverse people having assimilated captives from many tribes. The nineteenth century Oklahoma Seneca also were a mixed band which was further enriched in the 1870s by immigrants from Six Nations Reserve, Cattaraugus Reservation and elsewhere in New York. "These emigrations established family connections with the East . . . and resulted in significant cultural influences" (Sturtevant in Trigger 1978:539).

The nineteenth century Iroquois, thus, continue to share many effigy motifs. The data base of seventeenth and eighteenth century Seneca ladles permits an evaluation of the change seen in nineteenth century Seneca ladles which is not possible for the other Iroquois without additional data.

Plate 17. Ladle, wood. (192) *Ethnological.* **Seneca Iroquois, Tonawanda Reservation, c. nineteenth century. Museum of the American Indian (14/5688). About one and one quarter times actual size. Drawing by Gene Mackay.**

Plate 18. Ladle, wood. (200) *Ethnological.* Seneca Iroquois, Cattaraugus/Tonawanda Reservations, c. nineteenth century. Museum of the American Indian (2/9611). About one and one quarter times actual size. Drawing by Gene Mackay.

Plate 19. Ladle, wood. (231) *Ethnological.* Seneca Iroquois, c.A.D. 1846. RMSC (Lewis Henry Morgan, 70.89.37). About one and one-quarter times actual size. Drawing by Gene Mackay.

Plate 20. Ladle effigy, wood. (232) *Ethnological.* Seneca Iroquois, late nineteenth century? RMSC AE 1162. About two times actual size. Drawing by Gene Mackay.

Plate 21. Ladle, wood. (258) *Ethnological.* Seneca Iroquois, c.A.D. 1846. RMSC (Lewis Henry Morgan, 70.89.39A). About one and one-quarter times actual size. Drawing by Gene Mackay.

Plate 22. Ladle, wood. (302) *Ethnological.* Cayuga Iroquois, Grand River Reservation, c. nineteenth century. Museum of the American Indian (2/4417). About one and one quarter times actual size. Drawing by Gene Mackay.

Plate 23. Ladle, effigy, wood. (320) *Ethnological.* Onondaga Iroquois, late nineteenth century? American Museum of Natural History (50/7239). About one and one quarter times actual size. Drawing by Gene Mackay.

85

Plate 24. Ladle, wood. (361) *Ethnological.* Mohawk Iroquois, early nineteenth century. RMSC AE 7184. About one and one-quarter times actual size. Drawing by Gene Mackay.

Plate 25. Ladle effigy, wood. (366a) *Ethnological.* Mohawk Iroquois, late nineteenth century? American Museum of Natural History (50.1/1555). About one and one quarter times actual size. Drawing by Gene Mackay.

Plate 26. Ladle, effigy, wood. (450) *Ethnological.* **Iroquois, c.A.D. 1825–50. RMSC 6161/177. About three times actual size. Drawing by Gene Mackay.**

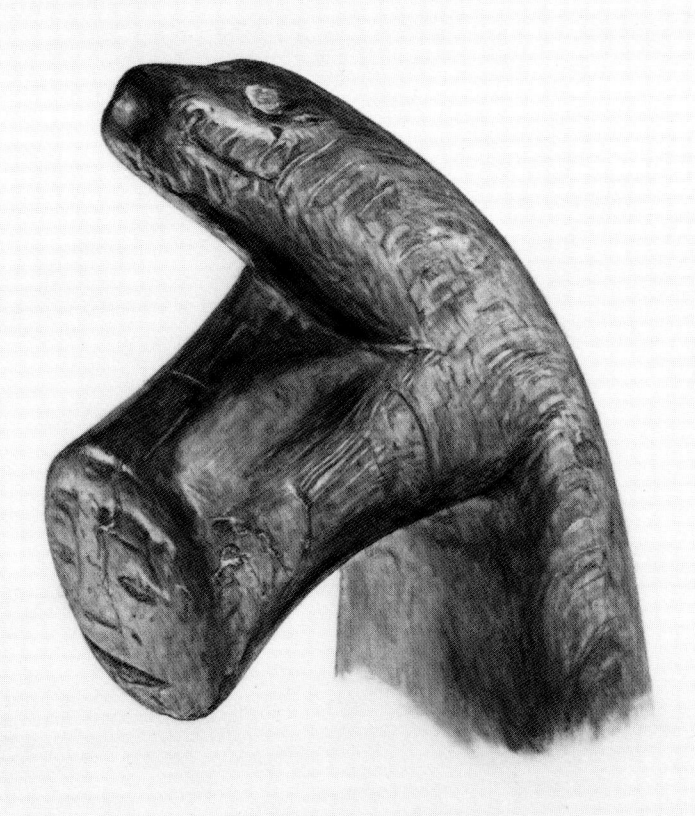

CHAPTER V

CHANGES IN EFFIGY MOTIFS

Among the Seneca, many effigy subjects which were common in the seventeenth and eighteenth centuries, such as human heads, clenched fists, reclining figures, turtle, snake/eel, panther, beaver and otter, are not present in the nineteenth century sample. (It is acknowledged that the amount of bias in the sample is unknown.) The Seneca ladles which have been preserved exhibit such traditional themes as the long-billed bird (formerly heron, now snipe), owl, swan/goose, bear/wolf and the combination of bear and human. Retention of bear/wolf, owl and swan/goose effigies may be evidence of the continuing importance of mystic societies. Only bear/wolf and snipe survive among clan motifs (probably an artifact of differential preservation or collection). More importance now seems to be attached to mystic animals than to mystically powerful humans.

A major nineteenth century modification is the use of the human figure in situational poses. The human effigies no longer are suggestive of magic and power. Instead, they portray such scenes as a human pulling on a boot, a human eating from a bowl, acrobats, wrestlers and humans holding placards. (The human figure in situational poses is also found in nineteenth century Cayuga, Onondaga and Mohawk ladles. The Mohawk ladle (366a) shows a seated human figure. It is illustrated in Plate 25.)

Another noticeable change in nineteenth century Seneca ladle effigies is the lack of specificity of the bird or animal being carved. In the seventeenth and eighteenth centuries, bears, hawks, panthers, beaver, etc. had identifiable "field marks", a flat, cross-hatched beaver tail, a long panther tail, a hooked raptor beak, or a flattened bear snout. By contrast, the nineteenth century sample shows generalized bird heads, generalized animals and a large number of generalized passerines. This does not stem from lack of carving ability as is evidenced by the recognizable owl, swan, buffalo, cat and squirrel effigies.

CHANGES IN OTHER ATTRIBUTES

The changes which can be observed in other attributes of the wooden ladles, such as the handle with back hook and the bowl which in the nineteenth century becomes on the average more wide than long (an average ratio of .68 as compared with 1.0 and .82 for the seventeenth and eighteenth centuries, respectively), may be attributable to European influence. Settlers from European dairying communities brought with them the tradition of butter making with its accompanying equipment, including wooden butter paddles and cream skimmers with back hooks to keep them from slipping into the milk. Dairy cattle were reported in Quebec as early as 1623 (Wrong 1939:50) and in New Netherlands by 1625 (Jameson 1909:82).

The introduction of the flat bottom brass trade kettle, which eventually replaced the round bottom native-made pottery, may have influenced the shape of the ladle bowl, which, particularly among the Seneca, became more wide than long (and, effectively, a better scraper). The shape of the bowl of the nineteenth century Seneca ladles thus can be seen as a functional adaptation rather than a European imitation. By contrast, the spoons of the Southeastern Indians continued to be "more pointed and triangular" than either European or Iroquoian types (Sturtevant 1979:199). European metal spoons imported into the colonies in the seventeenth and eighteenth centuries were typically of the "apostle" type, topped by an effigy of a saint, and bearing a bowl either fig-shaped or egg-shaped (Rainwater 1976:34). In the Rochester Museum Collections, metal spoons with apostle, pine knop and plain finials have been preserved from seventeenth century Seneca sites.

RELIGIOUS INFLUENCES ON THE SENECA

By the end of the eighteenth century, the Seneca Iroquois and the Europeans had been in contact for approximately 250 years. For most of that period the Indians had more than held their own in what ultimately would be an unequal struggle. By the end of the sixteenth century, European trade goods accounted for less than twenty-five per cent of the material culture. This would increase to more than seventy-five per cent by 1700 (Wray & Schoff 1953).

In mid-seventeenth century, the Seneca eliminated as enemies the Wenro (1638), the Huron (1649), the Petun (1650), the Neutrals (1651), the Erie (1656) and finally the Susquehannocks (1675) (Fenton 1940). The Seneca were stung but not seriously hurt by the French & Indian retaliatory expedition in 1687 led by the then governor of Canada, the Marquis de Denonville.

Conflicting allegiances factionalized the Seneca during the eighteenth century, as the western group leaned to the French and the eastern to the British. It was the Revolutionary War between the British and Americans that contributed largely to the Seneca's marginal position at the end of the eighteenth century. Few of the Seneca had sided with the victorious Americans during the conflict, and thus, they had no firm ground from which to negotiate and subsequently ceded most of their land to the new nation, retaining a few reserves for themselves. During these years the traditional matrilineal clan system apparently had weakened, and traditional male and female roles were upset. Indeed, it had been a tumultuous two and a half centuries.

It seems likely that the changes observed in Seneca ladle effigies after the middle of the eighteenth century reflected the declining fortunes of the Seneca, and, by analogy, of the other Iroquois, and also their search for more efficacious symbols to replace those which seemed to have lost power.

Many influences had been brought to bear on the Seneca during the previous 200 years which could have affected the choices they made for new symbols.

Missionary influence was sporadic throughout the seventeenth and eighteenth centuries. The Jesuits were followed by the United Brethren, the Quakers, the Baptists and other Protestant sects. The Moravians "taught the Indians new handicrafts" (Wallace 1951:288).

The teachings of the Seneca prophet Handsome Lake were instrumental in leading the Seneca away from the ravages of alcohol and inducing them to accept Eurpoean farming practices as taught to them by the Quakers (Deardorff 1956:591). Undoubtedly the missionaries attempted to de-mythologize native beliefs, and Handsome Lake especially condemned the mystic societies. These, however, were reintegrated into the ceremonies of the Longhouse religion after Handsome Lake's death (Shimony 1961:97–98), testimonial to their vital importance to the Seneca and perhaps accounting for the continuation of mystic animal symbolism on the ladle effigies.

MEDICINE LADLES

The existence of a class of magic, medicine ladles cannot be demonstrated with certainty. Those nineteenth century ladles identified ethnologically as being associated with medicine do not appear to share any common denominator of size, shape or effigy design.

The proliferation in the nineteenth century of ladles with bird head effigy may offer a clue. In his report on the consecration of the medicine of the Little Water Medicine Society at its semi-annual ceremony, Beauchamp notes that the master of ceremonies apportions the medicine to those doctors who are present by filling a duck's bill (Beauchamp 1901:156). A ladle which reportedly is in the collection of the Madison County (New York) Historical Society may relate to this. (The ladle has not been examined for this study.) It is said to have an effigy of a duck's head and bill, with only the lower mandible present, thus presenting a spoon-like utensil. It is not known if a ritual use is recorded for this ladle.

By the same association, it is possible that the prominent beak on the antler ladle from the Marsh Site (63), the Onondaga specimen (330) and the example from the Cattaraugus Reservation Seneca (174), as well as other Iroquois generalized bird head ladles, indicate specimens with medicinal function.

The owl is a bird with mystic associations to the Seneca. Fear is a dominant theme associated with owls at Six Nations Reserve (Shimony 1961:234). Owls represent witches (Curtin & Hewitt 1918; Parker 1923; Wray 1964). Both Christian and Longhouse Religion adherents at Onondaga Reservation "have an uneasy feeling that a screech owl is an omen of impending danger . . . " (Hendry 1964:375). The owl ladles known from this study also probably have mystic associations, although there is no direct evidence that they are "medicine" ladles. Owl ladles were recovered from the Markham Site and the Canawaugus Site as well as from the nineteenth century Cattaraugus and Tonawanda Reservations.

Ladles which are documented ethnologically as having been used for medicine include two non-effigy ladles (535, 541) from the Potawatomi Algonquians, an eared animal head ladle (565) from the Winnebago Siouans (for "love medicine"), a Sauk & Fox (Algonquians) turtle effigy ladle (602) used in the "war bundle feast," and the Dakota Siouan frog effigy ladle used by the medicine man who "doctored under the tutelage of the frog" (MAI-HF 6/7939) (Densmore 1948:Plate VII).

A Seneca Iroquois ladle collected by Arthur C. Parker from the Allegany Reservation in 1923 was said to have been "used only to give medicine" (NYSM 37149). This ladle was not available for examination.

The medicine of the Little Water Medicine Society is said to be apportioned with "a miniature ladle that holds as much of the powder as can be held on the tip of the blade of a small penknife" (Parker 1908:157). By analogy, the two small archaeological ladles with hour-glass effigies from the Marsh Site (70) and the Snyder-McClure Site (94) may have been medicine spoons.

Three nineteenth century Iroquois ladles have combination effigies which suggest a ceremonial theme, probably indicating the continuation of the shaman/doctor tradition. It will be recalled that the eighteenth century Honeoye Site produced a ladle (103) with a combination human and bear effigy which was suggestive of the bearskin costume worn by seventeenth and eighteenth century native shaman/doctors. A nineteenth century ladle (228), identified only as Seneca, also is carved with a human and bear combination. The human face has an elongated chin, and the bear face is carved on top of the human head. It is possible that a double mask is intended, as if the elongated chin would serve as a hand grip for a False Face Society mask, and the human masker were clothed in a bear skin.

A nineteenth century ladle (273) collected from the Oklahoma Seneca is carved with a human and turtle combination. The turtle is on the back of the human head and could represent simply a clan affiliation. The human face, however, is carved with the elongated chin which may indicate a mask. (The False Face Society ceremonial paraphernalia includes turtle shell rattles.)

A third nineteenth century combination ladle (422), identified as Iroquois, shows an effigy of a small human figure climbing up the neck of a human head. The meaning of this is not known, although it may have reference to the aforementioned clinging person myths. It may be significant that the human head in this effigy also is carved with an elongated chin.

NON-CEREMONIAL LADLE EFFIGIES

Among the human effigy ladles of the nineteenth century Iroquois, several are noteworthy for their situational, non-ceremonial themes. Mention was made previously of the eighteenth century Mohawk Iroquois ladle (130) showing a seated human figure which was characterized as a "warrior" because of what appeared to be a roached headdress. It may thus qualify as an eighteenth century precedent for the nineteenth century "secular" motifs. It is also possible, however, that the headdress may represent the sharp flint comb or crest of *Tawiskaron,* the evil one of the primeval twins in Iroquoian mythological belief. Hewitt (1903:292-5) recorded a Mohawk Iroquois version of *Tawiskaron's* birth. Hamell (1980b:4) summarizes: "It was with this sharp comb or crest that *Tawiskaron* cut through his mother's armpit, emerging there and killing her, rather (than) being born in the normal way."

Effigies of seated humans or prone humans are known also from Cherokee Iroquoians and Plains Siouans on their smoking pipes of the seventeenth and eighteenth centuries. These are believed, however, to have cultic associations (Brasser 1977), although they might seem to represent secular themes.

The nineteenth century ladles with situational, apparently non-traditional, human themes, then, may represent a special class, and the possibility may be considered that they were commissioned or, at the least, that the subject matter was suggested to the carver. The evidence, however, would seem to suggest that those who were in a position to influence subject matter were those who were most interested in the traditional ways of the Iroquois.

Lewis Henry Morgan, the nineteenth century Rochester, New York lawyer whose collection contained several of the situational human effigy Iroquois ladles, actively supported the traditionalist faction among the Seneca Iroquois in their struggle to regain their land during the 1840s and 1850s. Among the ladles in the Morgan Collection are effigies of a pair of human figures balancing on the shoulders of another pair of human figures; an effigy of a pair of wrestlers; and a carving of two pairs of human figures, seated back to back with a fifth human figure arched above them, hands and feet each upon the head of one of the sitting figures. Morgan (1852) noted in his report to the New York Cabinet of Antiquities that some of the ladles "were obtained among the Senecas, and the residue of the Iroquois in Canada" (Morgan 1852:82).

Among the twentieth century Onondaga Iroquois, there is a "self conscious effort to perpetuate aboriginal patterns" in the carving tradition, to the extent that the Indians "depend on the ethnographic literature as a design source and accept a white man as authority on old customs (Hendry 1964:388). This may have been true also of the nineteenth century Iroquois.

PERCHING BIRD EFFIGY

Among the nineteenth century New York Seneca Iroquois ladles, the perching bird effigy is noteworthy by virtue of its frequency of appearance. Encountered on a ladle only once before (a 1791 ethnological specimen (136) from the Cattaraugus Reservation), the perching bird effigy accounts for 74 of 135 ladles of the nineteenth century New York Seneca, or 55%.

The incidence of the perching bird (PB) effigy on nineteenth century ladles of the other Iroquois in the sample is tabulated below:

	PB Effigy	Total
Cayuga	2	18
Onondaga	5	26
Oneida	1	16
Mohawk	0	7
New York Iroquois	9	19
Grand River Iroquois	7	19
Canada Iroquois	1	15
Iroquois	7	30
	32	150

It is probable that the smaller percentage of perching bird effigies among the nineteenth century other Iroquois (as compared with the New York Seneca) is a result of the smaller sample size of ladles from other Iroquois. It is still significantly represented as compared with an incidence of 4 examples in 231 ladles from the nineteenth century Northern and Eastern Algonquians and Great Lakes peoples. Thus, the perching bird effigy appears fundamentally to be an Iroquois type.

The perching bird effigy is of a generalized passerine and is unlike the recognizable hawk/eagle raptor, the long-necked swan/goose or the long-billed heron/crane effigies common on the seventeenth century and eighteenth century ladles.

In the following pages, an inquiry is made into possible symbolic meaning(s) which the perching bird may have had for the Iroquois.

Functionally, the perching bird effigy ladle uses the bird's tail to hold the ladle to the rim of the pot and can be understood as an extension of the handle. Although in most examples the body of the handle also has a backwards angle, the bird's tail provides a positive hook. In other nineteenth century ladles, frequently the handle is double-hooked to provide secure anchorage to the pot rim, while the effigy is conceived independently on top of the handle.

The perching bird effigies are not all alike. Some have blunted tails, others have blunted heads, and some are quite abstract. All, however, are conventionalized to the degree that no particular species is recognizable.

Two suggestions are made to account for the frequency of the perching bird effigy ladle in the nineteenth century:

1) a bias or preference on the part of the collector
2) consumer preference.

Although there are undoubtedly biases in the study sample due to accidents of preservation, the collectors represent museum-trained anthropologists and fieldworkers for whom a representative sample of ladles from each group presumably would have been the ideal (Fenton 1940:160-164).

The large number of perching bird ladles, however, may be an artifact of customer preference, so that the Reservation carvers repeatedly carved the same image to please the consumer. If, indeed, this is a reason for the multiplicity of perching bird effigies, then the question arises as to what special significance the perching bird had for the Iroquois.

An influence on the Iroquoian perching bird motif of the nineteenth century wooden ladles probably was the Pennsylvania "Dutch" distelfink (thistle finch) or the European goldfinch. A "small bird motif" in European devotional art had spread in the thirteenth through the fifteenth centuries from northern France and central Italy to Spain, Flanders, Westphalia, Bavaria, Bohemia and Russia (Friedmann 1946:4), and was carried to the New World in the eighteenth century by Lutherans, Moravians and other seekers of religious freedom, many of whom settled in New York and Pennsylvania, the latter being known as the Pennsylvania "Dutch."

In Europe, a complex symbolism surrounded the goldfinch. The bird "was connected in the popular mind with an augur of a supernatural type in connection with disease" (Friedmann 1946:2). It also represented the soul as opposed to the body. It was associated with the theme of resurrection and of recovery after serious illness and, through the red coloration in the plumage of the European goldfinch, it symbolized sacrifice (Friedmann 1946:8).

The antecedent for the Christian goldfinch symbol was the mythical Charadrius (a curing symbol as far back as the sixth century B.C.), originally the golden plover, believed to be able to cure jaundice by absorbing to itself the yellow color characteristic of the disease. If, however, the bird averted its head from the patient, the prognosis was death and was construed as a moralistic judgment in later Christian eras as to whether or not the patient was lost to sin when forsaken by Christ or saved when He looked upon him (Friedmann 1946:12).

How much and in what detail the symbolism of the goldfinch was conveyed to the Iroquois is not known, although the fervor of the missionaries has been documented (Loskiel 1794; Beauchamp 1916; Wallace 1958). Among the multiple purposes of the Moravian missionaries, "They taught the Indians new handicrafts" (Wallace 1951:292). Not only were there seventeenth century Iroquoian mission schools, but also church schools were begun early in the eighteenth century which enrolled both Indian children and the children of settlers (Klees 1950:290). A Mennonite school teacher, Christopher Dock, rewarded industrious students with cards on which he had painted a bird or flower (Klees 1950:293)—possibly a distelfink or tulip. The Lutheran minister Heinrich Melchior Muhlenberg married the daughter of Conrad Weiser. Weiser was "in fact if not in title Pennsylvania's ambassador extraordinary to the Iroquois" (Klees 1950:81). Certainly, then, there was opportunity for dissemination of the symbolism carried by the goldfinch.

Acceptance by the Iroquoians of the symbolism associated with the goldfinch would not have presented a problem, as the traditional beliefs of the Iroquois concerning birds carried similar symbolic meanings. Also, there was no problem with specific species identification, as the Pennsylvania Dutch distelfink motif is painted as a generalized passerine, just as the perching bird effigy ladles are of generalized birds.

The Iroquois may have seen a relationship between the Charadrius/goldfinch which prophesied death when it averted its head from the patient and the Iroquois folk tale of the magic pipe, which at the owner's command was lit by two pigeons, but when the pigeons "fail to perform their duty" to light the pipe, the owner died (Randle 1953:633).

When they carved the perching bird ladle effigies, the Iroquois could have had in mind the migratory passenger pigeon which formerly nested in Iroquois territory in vast numbers, providing them with a quantity of food, and which occasioned an annual dance and songs of thanksgiving (Curtin & Hewitt 1918:666). The story of the origin of the Pigeon Dance has found a place in the Iroquois creation myth (Hewitt 1903:304).

The red breast of the passenger pigeon cock bird may have become associated in the Iroquois' mind with the red color of the Christian symbol of sacrifice.

The perching bird effigies could have been intended to represent the mourning dove. In the belief of the seventeenth century Huron Iroquois, the soul turned into a "turtle-dove" (Fenton 1977:238; JR 1636:287). A similar belief prevailed among the Seneca (Morgan 1962:174). The Christian symbol of Christ as a dove is well known.

If further sanction were needed by the Iroquois to assist them in acceptance of the bird symbol, it was the Seneca prophet Handsome Lake himself who confirmed the necessity for continuing the Great Feather Dance of Thanksgiving as one of the four sacred rituals authorized by the Creator (Parker 1913:41).

"From birds came all the Indian songs and dances" says the Iroquois story teller (Curtin 1923:184), and, as Wallace Chafe (1961:156) points out, the Seneca Iroquois word for song "is cognate with the Mohawk word borrowed into English as orenda. In all probability the Seneca word also once meant 'supernatural power,' one of whose manifestations was song."

In view of the situation in which the late eighteenth and early nineteenth century Iroquois found themselves, diminished in number, stripped of their former power and influence and essentially confined to reservations, it seems reasonable to think that the symbolism associated with birds would have had particular appeal to them and may account for the apparent popularity of the effigy on the nineteenth century wooden ladles.

It is clear that the Iroquois were selective in their choices of effigies. The tulip and heart, for example, so common in Pennsylvania Dutch folk art, carried as much symbolic meaning to the Eurpeans as the small bird, but only the latter also had meaning for the Iroquois.

SUMMARY

A comparison of nineteenth century with seventeenth and eighteenth century Iroquois ladles shows many differences. Not only has the morphology changed (probably a functional response to changes from round-bottomed, no-slip earthen cookware to flat-bottomed, slippery metal kettles) but also and perhaps more significantly, the effigies have changed.

Among the Seneca Iroquois, the symbolic content of the ladle effigies appears to have undergone considerable transformation. It may be recalled that a preponderance of the seventeenth and eighteenth century ladle effigies portrayed either humans with supernatural aspects, clan animals and birds or mystic creatures. By contrast, the nineteenth century ladles treat only marginally mystic humans and clans but present a larger number of mystic animal effigies.

In the following tabulation, animals or birds which functioned dually as clan and mystic symbols have been allocated to the clan.

Seneca Effigies

	Mystic Human		Clan Animal/Bird		Mystic Animal		Total
	No.	%	No.	%	No.	%	No.
17th C.	18	38%	26	55%	3	7%	47
18th C.	3	27%	5	46%	3	27%	11
19th C.	2	11%	5	26%	12	63%	19

These figures probably reflect certain realities in the Seneca's nineteenth century society. In accounting for the decrease in effigies with supernatural human attributes, it is likely that missionaries downplayed the concept of mystic humans in the role of shaman/doctors. The small number of clan effigies could be a consequence of Handsome Lake's inveighing against the practice of witchcraft and preaching that husband and wife should form the social unit rather than the matrilineal clan with its consequent brittle marriages.

Conservatism has been noted as a continuing trait of the Iroquois (Shimony 1961). In view of this, it is significant that mystic societies apparently were re-integrated into Seneca society despite Handsome Lake's admonitions against them. Among the nineteenth century effigies, it is the mystically associated long-necked bird, long-tailed otter, curer-bear (from the clan list) and the owl of witchcraft which survive. In the face of a fragmented social organization, the community-based ceremonialism of the mystic societies could have provided needed psychological support.

Perhaps the most devastating effect of the arrival of the Europeans in the New World was the co-occurring appearance of unprecedented epidemics of disease which scourged the Indian communities. As the epidemics continued, the decimated Indian populations pursued a search for powerful amulets. It may be that the perching bird effigy which emerged in the nineteenth century served as a composite symbol incorporating the most powerful beliefs of both Indians and the Europeans.

In considering the multitude of symbolic meanings attached to a generalized bird figure by both the Indians and the Europeans, it seems possible that the perching bird effigy absorbed many of these meanings. For example, the European dove equates with the Iroquoian passenger pigeon or the charadrius/finch with the Iroquoian pigeon cock bird. As previously mentioned Waugh notes,

> "The handles of spoons are frequently carved with designs which are ornamental, totemistic, or in response to dreams, particularly those occurring during some indisposition or illness. The dreams are interpreted by a local seer or medical practitioner, who decides upon the design, also the kind of wood, the presentation of such dream-objects to the patient being necessitated to secure recovery. Failure in this respect is believed to be followed by continued illness and eventually by death" (Waugh 1916:68). Paraphrasing the Jesuits (JR 1656:267), Waugh continues, "The custom seems to have been based upon the belief that the soul can depart from the body and that satisfaction of its desires must be obtained to bring about its return." (Waugh 1916:68).

Attesting the importance of dreams in Iroquoian belief, Fenton notes "Dreams have played a dominant role in culture change for the Iroquois . . . Ritual procedures have been altered and new customs instituted because certain individuals have revealed dreams which the longhouse officers have considered important enough to actualize" (Fenton 1936:4).

Thus it may be (perhaps as a result of a dream) that the perching bird served as a synthesis of symbolic meanings, lending credence to the thesis that the Iroquois chose certain effigies which reinforced traditional beliefs. Some changes in effigy content can be explained as responses to the contemporary society, but it also seems clear that the Seneca retained the effigies which symbolized concepts of central importance to them.

COLLECTIONS

AMNH	American Museum of Natural History, Department of Anthropology, New York, New York.
BC	Bacone College, Muskogee, Oklahoma.
BM	The British Museum, Ethnography Department, London, England.
BEC	Buffalo and Erie County Historical Society, Buffalo, New York.
BMS	Buffalo Museum of Science, Buffalo, New York.
DIA	Detroit Institute of Arts, Detroit, Michigan.
FMNH	Field Museum of Natural History, Department of Anthropology, Chicago, Illinois.
LSPM	Letchworth State Park Museum, Genesee State Park & Recreation Commission, Castile, New York.
MPM	Milwaukee Public Museum, Milwaukee, Wisconsin.
MAI-HF	Museum of the American Indian—Heye Foundation, New York, New York.
MH-P	Musee de l'Homme, Paris, France.
MV-B	Museum fuer Voelkerkunde, Berlin, West Germany.
NM-D	Nationalmuseet, Copenhagen, Denmark.
NMM	National Museum of Man, Ottawa, Canada.
NYSM	New York State Museum, Archaeology and Anthropology Department, Albany, New York.
OCHS	Ontario County Historical Society, Canandaigua, New York.
PM-HU	Peabody Museum of Archaeology & Ethnology, Harvard University, Cambridge, Massachusetts.
PM-S	Peabody Museum of Salem, Salem, Massachusetts.
PM-YU	Peabody Museum of Natural History, Yale University, New Haven, Connecticut.
PRM	Pitt Rivers Museum, Department of Ethnology and Prehistory, Oxford, England.
RHS-RMSC	Rochester Historical Society, Rochester Museum & Science Center, Anthropology Section, Rochester, New York.
RMSC	Rochester Museum & Science Center, Anthropology Section, Rochester, New York.
ROM	Royal Ontario Museum, Department of New World Archaeology, Toronto, Canada.

Ladles in the study sample which were taken from illustrations but whose present location was not verified are listed below by citation of author and publication.

Abrams, George
    1965 The Cornplanter Cemetery. *Pennsylvania Archaeologist* 35 (2): 70, fig. 3a.

Beauchamp, William M.
    1905 Aboriginal Use of Wood in New York. *New York State Museum Bulletin* 89 (Plate 18, fig. 92).

Densmore, Frances
    1929 Chippewa Customs. *Bureau of American Ethnology Bulletin* 86 (Plate 17). Smithsonian Institution, Washington, D.C.

Maurer, Evan N.
    1977 *The Native American Heritage. A Survey of North American Indian Art.* The Art Institute of Chicago. University of Nebraska Press, Lincoln, 127, fig. 137.

Ritzenthaler, Robert
    1976 Woodland Sculpture. *American Indian Art* Autumn 40, fig. 12.

Speck, Frank G.
    1945 The Iroquois: A Study in Cultural Evolution. *Cranbrook Institute of Science Bulletin* 23:84. Bloomfield Hills, Michigan.

APPENDIX

Text No.	Tribe	Location	Effigy	Notes	Coll.	Cat. No.
1	Seneca	Cameron Site	—	—	RMSC	5443/41
2	"	"	—	—	"	5067/41
3	"	Factory Hollow	—	Perforated	"	5136/102
4	"	"	Abstract	—	"	38/102
5	"	"	Bear/Wolf	—	"	5005/102
6	"	"	—	Top Broken	"	44/102
7	"	"	—	"	"	5004/102
8	"	"	—	"	"	6088/102
9	"	"	—	"	"	61/102
10	"	Warren Site	Human	Bowl	"	40/89
11	"	"	Hawk/Eagle	—	"	40/89
12	"	"	Bear/Panther	—	"	40/89
13	"	Steele Site	Human	Antler	"	5000/100
14	"	"	"	Blue Color	"	552/100
15	"	"	"	—	"	169/100
16	"	"	Clenched Fist	—	"	522/100
17	"	"	"	—	"	604/100
18	"	"	Bear/Wolf	—	"	546/100
19	"	"	Horse	—	"	547/100
20	"	"	Turtle	—	"	6079/100
21	"	"	"	—	"	6222/100
22	"	"	Snake	—	"	437/100
23	"	"	—	—	"	6469/100
24	"	"	—	Perforated	"	605/100
25	"	"	—	Top Broken	"	545/100
26	"	"	—	"	"	6117/100
27	"	Power House	—	Toggle	"	1186/24
28	"	"	Bear/Wolf	—	"	AR 29201
29	"	"	Bear	—	"	AR 42684
30	"	"	Otter	Antler	"	1034/24
31	"	"	—	—	"	3402/24
32	"	"	—	—	"	1185/24
33	"	"	—	Top Broken	"	216/24
34	"	Dann Site	Human	—	"	1123/28
35	"	"	Clenched Fist	Perforated	"	5636/28
36	"	"	Abstract	—	"	187/28
37	"	"	"	—	"	6190/28
38	"	"	Reclining Fig.	Antler	"	6002/28
39	"	"	Beaver	—	NYSM	21139
40	"	"	"	—	RMSC	1083/28
41	"	"	Turtle	Pipe Frag.?	"	205/28
42	"	"	—	Peforated	"	2091/28
43	"	"	—	—	"	6053/28
44	"	"	—	—	MAI-HF	22/2957
45	"	"	—	Top Broken, Antler	NYSM	21133
46	"	"	—	"	RMSC	3167/28
47	"	"	—	"	"	6023/28
48	"	"	—	"	"	6022/28
49	"	"	—	"	"	181/28
50	"	"	—	"	"	6106/28
51	"	Markham Site	Owl	—	"	145/T
52	"	Roch. Junction	Human	Antler	"	664/29
53	"	"	Hawk/Eagle	—	"	255/29
54	"	"	Bear	Antler, also engraved	"	5034/29
55	"	"	—	Top Broken	"	309/29

Text No.	Tribe	Location	Effigy	Notes	Coll.	Cat. No.
56	"	Marsh Site	Human	Pregnant Female	"	807/99
57	"	"	"	Antler	OCHS	I 279
58	"	"	"	—	RMSC	5047/99
59	"	"	"	Female	"	1783/99
60	"	"	Clenched Fist	—	"	420/99
61	"	"	Abstract	—	"	853/99
62	"	"	Reclining Fig.	—	"	1938/99
63	"	"	Heron/Crane	Antler	OCHS	I 278
64	"	"	Hawk/Eagle	—	RMSC	835/99
65	"	"	Bears	Two	"	AR 18401
66	"	"	Bear/Wolf	—	OCHS	I 282
67	"	"	Bear/Wolf	—	"	I 280
68	"	"	Panther	Also hourglass	RMSC	5033/99
69	"	"	Turtle	—	"	836/99
70	"	"	Hourglass	—	"	/99
71	"	"	—	Perforated	"	1848/99
72	"	"	—	Top Broken	"	1784/99
73	"	"	—	"	"	421/99
74	"	"	—	"	"	5048/99
75	"	Boughton Hill	Reclining Fig.	—	"	2284/103
76	"	"	Abstract	—	NYSM	—
77	"	"	Bear/Wolf	Antler	RMSC	2241/103
78	"	"	"	—	"	28/103
79	"	"	"	—	"	2287/103
80	"	"	Turtle	—	"	2346/103
81	"	"	"	—	"	5007/103
82	"	"	"	Shell	"	2212/103
83	"	"	Abstract	Antler	NYSM	—
84	"	"	—	Top Broken	"	—
85	"	"	—	"	"	—
86	"	"	—	"	RMSC	2323/103
87	"	"	—	"	"	2236/103
88	"	"	—	"	"	2322/103
89	"	Beal Site	Animal Head	—	"	135/98
90	"	"	—	Top Broken	"	293A/98
91	"	Snyder-McClure	Reclining Fig.	—	"	AR 18403
92	"	"	Animal Head	—	"	AR 18402
93	"	"	Turtle	Pipe Frag. &	"	5003/132
94	"	"	Hourglass	Antler	"	AR 18589
95	"	"	Abstract	Perforated	"	5002/132
96	"	"	—	Top Broken	"	—
97	"	"	—	"	"	—
98	"	"	—	—	"	AR 18590
99	"	Townley Read	—	", Antler	"	5/160
100	"	Huntoon Site	Human	Red Color; Maskette?	"	85/159
101	"	"	Abstract	Human Buttocks?	"	6160/159
102	"	"	—	Top Broken	"	6137/159
103	"	Honeoye Site	Human/Bear	—	"	117/118
104	"	"	—	Top Broken	"	118/118
105	"	Kendaia Site	Swan/Goose	—	"	3/201
106	"	Fall Brook	Beaver	—	"	AR 29331
107	"	Fall Brook Site	—	Top Broken	"	254/W
108	"	Big Tree Site	Hawk/Eagle	—	"	104/178
109	"	"	Otters	Two	"	51/178
110	"	Canawaugus Site	Owl	—	"	AR 29836
111	"	"	Bear/Wolf	—	"	AR 29851
112	Onondaga	Jayne-LaPoint	Human	—	PrivColl	—
113	"	Pen Site	Human	—	"	—
114	Cayuga	Rogers Site	—	Perforated	RMSC	AR 29042
115	Neutral	Grimsby Site	Human	—	ROM	—
116	"	St. David's Site	Snake?	Antler	MAI-HF	—
117	"	Lake Medad Site	—	—	"	—
118	"	Grimsby Site	Weasel	—	ROM	—
119	"	"	—	Antler	"	—
120	"	"	—	—	"	—

Text No.	Tribe	Location	Effigy	Notes	Coll.	Cat. No.
121	Wyandot	Upper Sandusky	Human	—	MAI-HF	14/9600
122	Potawatomi	Michigan	Hand	—	DIA	51.10
123	Chippewa	Rainy Lake Site	—	—	ROM	—
124	"	"	Birds	Two	"	—
125	Mahican	New York	Animal	—	NYSM	36915
126	"	"	—	—	MAI-HF	19/8131
127	Delaware	Connecticut	—	—	"	19/8360
128	Onondaga	Coye Site	Horse	—	PrivColl	—
129	"	Sevier Site	Beaver	Bowl	"	—
130	Mohawk	Mohawk Valley	Human	—	BC	—
131	Onondaga	Coye Site	Human	—	PrivColl	—
132	"	Sevier Site	—	Top Broken	"	—
133	"	Coye Site	Bird	—	"	—
134	Hist. Iroq.	Genoa Fort Site	Hourglass	Antler	RMSC	6069/205
135	Ottawa	Gros Cap Site	Birds	Two, Antler	FMC	—
136	Seneca	Cattaraugus Res.	Bird	18th Century	MAI-HF	16/2700
137	"	"	Human	19th C.	"	6/376
138	"	"	"	—	"	6/1132
139	"	"	"	—	"	20/1868
140	"	"	Swan/Goose	—	"	2/1059A
141	"	"	Owl	—	PM-HU	62624
142	"	"	Bird	—	MAI-HF	6/383
143	"	"	"	—	PM-HU	65436
144	"	"	"	—	NMM	III-I-15
145	"	"	"	—	MAI-HF	1716
146	"	"	"	—	"	1720
147	"	"	"	—	"	1724
148	"	"	"	—	"	2269
149	"	"	"	—	"	2270
150	"	"	"	—	"	2/1059B
151	"	"	"	—	"	2/1059C
152	"	"	"	—	"	6/378
153	"	"	"	—	"	6/1129
154	"	"	"	—	"	8/7830
155	"	"	"	—	"	14/4982
156	"	"	"	—	"	16/2699
157	"	"	"	—	"	20/6725
158	"	"	"	—	"	20/6726
159	"	"	"	—	RMSC	AE 14
160	"	"	"	—	PM-HU	62627
161	"	"	"	—	"	62625
162	"	"	"	—	MAI-HF	1714
163	"	"	"	—	BMS	C5279
164	"	"	"	—	MAI-HF	1881
165	"	"	"	—	"	9461B
166	"	"	"	—	"	8/7829
167	Seneca	Cattaraugus Res.	Bird	—	"	20/6728
168	"	"	"	—	"	20/6727
169	"	"	"	—	"	24/1336
170	"	"	"	—	"	8/7831
171	"	"	"	—	"	1718
172	"	"	"	—	"	8/7828
173	"	"	Bird Head	—	"	23/7823
174	"	"	"	—	NYSM	36736
175	"	"	Man & Buffalo	—	MAI-HF	6/1131
176	"	"	Animal Heads	Double	"	20/5538
177	"	"	Animal Head	—	"	22/4275
178	"	"	"	—	"	23/7822
179	"	"	"	—	NMM	III-I-17
180	"	"	Buffalo	—	MAI-HF	6/377
181	"	"	—	Perforated	"	1725

Text No.	Tribe	Location	Effigy	Notes	Coll.	Cat. No.
182	"	"	—	—	PM-HU	62626
183	"	"	—	—	"	65437
184	"	"	—	—	MAI-HF	1709
185	"	"	—	—	"	1719
186	"	"	—	—	"	9461A
187	"	"	—	—	"	9461C
188	"	"	—	—	"	24/7821
189	"	"	—	—	"	24/1304
190	"	"	—	—	"	20/1867
191	"	"	—	—	"	20/1865
192	"	Tonawanda Res.	Human & Boot	—	"	14/5688
193	"	"	Human & Dog	—	NYSM	37000
194	"	"	Bird	—	NMM	III-I-757
195	"	"	Bird	—	MAI-HF	22/4258B
196	"	"	Animal	—	"	1876
197	"	"	Cat	—	"	24/1334
198	"	"	Horse	—	"	22/4258C
199	"	"	—	Perforated	"	22/4258A
200	Seneca	Catt. or Ton. Res.	Human	—	"	2/9611P
201	"	"	Snipe/Heron	—	"	2/9611E
202	"	"	Bird, owl?	—	"	2/9611Q
203	"	"	Bird	—	"	2/9611N
204	"	"	"	—	"	2/9611L
205	"	"	"	—	"	2/9611A
206	"	"	"	—	"	2/9611C
207	"	"	"	—	"	2/9611F
208	"	"	"	—	"	2/9611D
209	"	"	"	—	"	2/9611I
210	"	"	"	—	"	2/9611J
211	"	"	"	—	"	2/9611O
212	"	"	"	—	"	2/9611T
213	"	"	"	—	"	2/9611K
214	"	"	"	—	"	2/9611S
215	"	"	Fox/Wolf	—	"	2/9611B
216	"	"	Ball/Disc	Perforated	"	2/9611M
217	"	"	—	—	"	2/9611R
218	"	"	—	—	"	2/9611G
219	"	"	—	—	"	2/9611H
220	"	Allegany Res.	Bird	—	RMSC	AE 499
221	"	"	"	—	PM-YU	25299
222	"	"	"	—	MAI-HF	22/4285
223	"	"	"	—	RMSC	AE 335
224	"	"	"	—	"	AE 340
225	"	"	"	—	Abrams	—
226	"	"	—	—	MAI-HF	13/6634
227	"	—	Human	Kneeling, on all fours	BMS	C198
228	"	—	Human/Bear	Double Mask	AMNH	50.1/1513
229	"	—	Acrobats	LHM Coll.	RMSC	70.89.40
230	"	—	Wrestlers	"	NYSM	LHM-186
231	"	—	Leg & Foot	"	RMSC	70.89.37
232	"	—	Swan/Goose	—	"	AE 1162
233	"	—	"	—	MAI-HF	8957M
234	Seneca	—	Bird	—	"	8957F
235	"	—	"	—	AMNH	50/6197
236	"	—	"	—	"	50/6205
237	"	—	"	—	"	50/5548
238	"	—	"	—	RMSC	AE 411
239	"	—	"	—	NMM	III-I-1365
240	"	—	"	—	LSPM	—
241	"	—	"	—	AMNH	50/6204
242	"	—	"	—	MAI-HF	8957Q
243	"	—	"	—	AMNH	50.1/1516
244	"	—	"	—	MAI-HF	8957H

Text No.	Tribe	Location	Effigy	Notes	Coll.	Cat. No.
245	"	—	"	—	"	8957G
246	"	—	"	—	"	8957O
247	"	—	"	—	AMNH	50/6198
248	"	—	"	—	MAI-HF	8957J
249	"	—	"	—	"	8957L
250	"	—	"	—	"	8957K
251	"	—	"	—	"	8957N
252	"	—	"	—	"	8957A
253	"	—	"	—	"	8957C
254	"	—	"	—	"	8957E
255	"	—	"	Jemison Ladle?	LSPM	—
256	"	—	"	LHM Coll.	RMSC	70.89.38
257	"	—	"	LHM Coll.	"	70.89.39
258	"	—	"	LHM Coll.	"	70.89.39A
259	"	—	"	—	MAI-HF	6/7700
260	"	—	"	—	"	8957P
261	"	—	Bear/Wolf	Double	AMNH	50/6206
262	"	—	Squirrel	LHM Coll.	NYSM	36818
263	"	—	Frog?	—	MAI-HF	2/9924
264	"	—	Ball	—	RMSC	AE 388
265	"	—	—	—	MAI-HF	8957D
266	"	—	—	—	"	16/9627
267	Seneca	—	—	—	AMNH	50/6196B
268	"	—	—	—	RMSC	AE 134
269	"	—	—	—	MAI-HF	8957I
270	"	—	—	—	"	8957B
271	"	—	—	—	AMNH	50.1/1519
272	"	Oklahoma	Human	Painted, Pencilled	NMM	III-I-233
273	"	"	Human&Turtle	—	MAI-HF	16/9377
274	"	"	Bird	—	"	16/9381
275	"	"	"	—	"	16/9383
276	"	"	Swan/Goose	—	"	16/9382
277	"	"	"	—	"	16/9380
278	"	"	"	—	"	16/9379
279	"	"	Bird Head	—	"	16/9384
280	"	"	"	—	"	16/9386
281	"	"	"	—	"	1/9633
282	"	"	Bear	Incised, Painted	NMM	III-I-235
283	"	"	Dog/Wolf	Incised	"	III-I-234
284	"	"	Bear/Wolf	—	MAI-HF	16/9388
285	"	"	Otter	—	"	16/9376
286	"	"	Buffalo	—	"	16/9378
287	"	"	Bear/Wolf Head	—	"	20/7279
288	"	"	Animal Head	"C" Shape	"	16/9387
289	"	"	"	—	"	1/9630
290	"	"	—	Perforated	NMM	III-I-430
291	"	"	—	Perforated	MAI-HF	16/9369
292	"	"	—	—	"	1/9631
293	"	"	—	—	"	16/9375
294	"	"	—	—	"	16/9370
295	"	"	—	—	"	16/9374
296	"	"	—	—	"	20/7277
297	"	"	—	—	"	20/7276
298	Iroquois	"	—	Perforated	NMM	III-I-237
299	"	"	—	—	"	III-I-433
300	Cayuga	Cattaraugus Res.	Human	—	NYSM	37124
301	"	—	Human	—	"	36960
302	"	Grand River Res.	Ducks	Three, & Engraved	MAI-HF	2/4417
303	"	Six Nations Res.	"	Three, & Engraved	AMNH	50.1/1690
304	"	Grand River Res.	"	Two, & Engraved	MAI-HF	2/3387
305	"	"	Bird	—	"	18/4735
306	"	Six Nations Res.	Duck	—	NMM	III-I-610

Text No.	Tribe	Location	Effigy	Notes	Coll.	Cat. No.
307	"	"	Bird	—	"	III-I-273
308	"	"	Bird Head	—	PM-S	E26304
309	"	"	Turtle	Perforated	"	E26303
310	"	"	Snake	—	NMM	III-I-612
311	"	Oklahoma	—	Perforated	"	III-I-431
312	"	"	—	—	"	III-I-432
313	"	Grand River Res.	—	—	MAI-HF	16/2018
314	"	Six Nations Res.	—	—	PM-S	E26302
315	"	"	—	—	NMM	III-I-613
316	"	"	—	—	"	III-I-897
317	"	"	—	—	"	III-I-897B
318	Onondaga	New York	Man & Book	—	FMNH	92093
319	"	Onondaga Castle	Human	Uplifted Face	AMNH	50/7235
320	"	"	"	Backbend	"	50/7239
321	"	"	"	Reclining w/Jug	NYSM	50070
322	"	Onondaga Res.	Swan/Goose	—	"	50075
323	"	New York	"	—	FMNH	92089
324	"	Onondaga Res.	Owl	—	NYSM	50073
325	"	New York	Bird	—	"	50077
326	"	Onondaga Res.	"	—	"	50071
327	"	—	"	—	MAI-HF	22/4246
328	"	—	"	—	PM-HU	62811
329	"	—	"	—	Beauchamp 1905: pl. 18, #9	
330	"	Onondaga Res.	Bird Head	—	NYSM	50069
331	"	Onondaga Castle	"	—	AMNH	50/7237
332	"	—	Animal & Swan	—	Beauchamp 1905: pl. 21, #10	
333	"	New York	"	—	FMNH	92091
334	Onondaga	Onondaga Res.	Sheep	—	NYSM	50072
335	"	"	Buffalo	—	MAI-HF	20/1247
336	"	New York	Rabbit	—	FMNH	92092
337	"	—	Bear	—	MAI-HF	20/1248
338	"	New York	Animal	—	"	16/9452
339	"	—	Turtle	—	RMSC	AE 2892
340	"	Onondaga Castle	Ball	Enclosed Rattle	AMNH	50/7238
341	"	—	Abstract	Banana Shape	NYSM	50076
342	"	New York	—	—	MAI-HF	21/3213
343	"	New York	—	—	FMNH	92090
344	Oneida	Six Nations Res.	Human	—	AMNH	50.1/1801
345	"	Wisconsin	Human	—	MPM	5925
346	"	Six Nations Res.	Clenched Fist	—	MAI-HF	1/4624
347	"	New York Res.	Bird	—	"	24/1402
348	"	Wisconsin	Swan/Goose	—	Speck 1945:84	
349	"	Wisconsin	Bird	Grouse	Speck 1945	
350	"	"	"	"	Speck 1945	
351	"	"	"	Woodcock	Speck 1945	
352	"	Six Nations Res.	Bear	—	AMNH	50.1/1797
353	"	Wisconsin	Squirrel	—	Speck 1945	
354	"	Six Nations Res.	—	—	NMM	III-I-274
355	"	Oneidatown, Ont.	—	—	"	III-I-796
356	"	Wisconsin	—	—	MPM	5931
357	"	"	—	—	"	5925
358	"	"	—	—	"	5922
359	"	"	—	—	"	5923
360	Mohawk	Bay of Quinte	—	Animal	MAI-HF	1/4619
361	"	—	Beaver	—	RMSC	AE 7184

Text No.	Tribe	Location	Effigy	Notes	Coll.	Cat. No.
362	"	St. Regis	—	—	Beauchamp	1905
363	"	Caughnawaga Res.	—	—	MAI-HF	1/2815
364	"	"	—	—	NMM	III-I-995
365	"	Six Nations Res.	—	—	MAI-HF	4/419
366	"	"	—	—	FMNH	55782
366a	"	—	Human	—	AMNH	50.1/1555
367	Tuscarora	Six Nations Res.	—	—	NMM	III-I-51
368	Iroquois	New York	Swan/Goose	—	MPM	24120
369	"	"	Cock/Comb	—	"	24136
370	"	"	Bird	—	"	24146
371	"	"	"	—	"	24145
372	"	"	"	—	"	2414?
373	"	"	"	—	"	24141
374	"	"	"	—	"	24140
375	"	"	"	—	"	24139
376	"	"	"	—	"	24138
377	"	"	"	—	"	24137
378	"	"	"	—	"	24127A
379	"	"	—	—	"	24122
380	"	"	—	—	"	24123
381	"	"	—	—	"	24127B
382	"	"	—	—	"	24144
383	"	"	—	—	PM-S	16406
384	"	"	—	—	MH-P	85.78.186
385	"	"	Abstract	Animal or Human?	MPM	24124
386	"	"	Abstract	Scalloped Head?	"	24117
387	"	Six Nations Res.	Clenched Fist	—	NMM	III-I-1066
388	"	"	Bird	—	"	III-I-358
389	"	"	"	—	"	III-I-360
390	"	"	"	—	MAI-HF	8289
391	"	"	"	—	AMNH	50.1/1869
392	"	"	"	—	MV-B	B12686
393	"	"	"	—	NMM	III-I-357
394	"	"	"	—	AMNH	50.1/1870
395	"	"	Bird Head	—	NMM	III-I-614
396	"	"	"	—	"	III-I-1065
397	"	"	Bear/Wolf	—	"	III-I-1064
398	"	"	Cat	—	AMNH	50.1/1881
399	"	"	Snake	—	FMNH	55781
400	"	"	—	Chip carved	"	52647
401	Iroquois	Six Nations Res.	—	—	NMM	III-I-355
402	"	"	—	—	"	III-I-356
403	"	"	—	—	"	III-I-354
404	"	"	—	—	"	III-I-611
405	"	Canada	Swan/Goose	—	MPM	24548
406	"	"	"	—	"	24546
407	"	"	"	—	"	24551
408	"	"	Bird	—	"	24545
409	"	"	Buffalo	—	"	24543
410	"	"	Animals	Two	"	24541
411	"	"	Animals	Two	"	24547
412	"	"	Abstract	—	"	24554
413	"	"	—	—	"	24538
414	"	"	—	—	"	24540
415	"	"	—	—	"	24549
416	"	"	—	—	"	24542
417	"	"	—	—	"	24550
418	"	"	—	—	"	3370
419	"	"	—	—	"	3371
420	"	—	Human	Holding Object	PM-S	30727
421	"	—	Human	Arched Back	AMNH	50/6737

Text No.	Tribe	Location	Effigy	Notes	Coll.	Cat. No.
422	"	—	Human & Human Mask?	—	"	50/6736
423	"	—	Owl	—	"	
424	"	—	Bird	—	MAI-HF	22/7648
425	"	—	"	—	BEC	66-149
426	"	—	"	—	"	66-146
427	"	—	"	—	MPM	21131
428	"	—	"	—	NM-D	HC 377
429	"	—	"	—	AMNH	50/6745
430	"	—	"	—	RHS-RMSC	T-186
431	"	—	Bird Head	—	BEC	66-167
432	"	—	Swan/Goose	—	AMNH	50/6742
433	"	—	"	—	"	50/6743
434	"	—	Swan/Goose	—	"	50/6744
435	"	—	Swan/Goose	—	"	
436	"	—	Bear/Wolf	—	"	50/6738
437	"	—	Buffalo	—	"	50/6740
438	"	Caughnawaga	Rabbit	—	NMM	III-I-735
439	"	—	Horse	—	AMNH	50/6741
440	"	Caughnawaga	Bear/Wolf Head	—	NMM	III-I-736
441	"	—	Animal	—	AMNH	50/6739
442	"	—	Ball/Disc	—	BEC	66-150
443	"	—	Abstract	—	AMNH	50/6735
444	"	Caughnawaga	—	—	NMM	III-I-734
445	"	—	—	—	MAI-HF	22/4668
446	"	—	—	—	BEC	66-163
447	"	—	—	—	AMNH	50/6746
448	"	—	—	—	"	50/6750
449	"	—	—	—	BEC	66-161
450	Probably Iroquois		Human & Salamander	Mask?	RMSC	6161/177
451	"	"		Bird	—	"	6160/177
452	"	"	—	Bird	—	"	AE 2893
453	"	"	—	"	—	NYSM	36825
454	"	"	—	"	—	"	37114
455	"	"	—	"	-	BEC	66-122
456	"	"	—	"	—	NYSM	37119
457	"	"	—	"	—	"	36291
458	"	"	—	"	—	"	37130
459	"	"	—	"	—	"	37120
460	"	"	—	"	—	"	37123
461	"	"	—	"	—	"	37116
462	"	"	—	"	—	"	37131
463	"	"	—	Bird Head	—	BMS	C 199
464	"	"	—	"		LSPM	
465	"	"	—	Animal Head	—	RMSC	6159/177
466	"	"	—	"	—	NYSM	37122
467	"	"	—	"	—	"	37110
468	Delaware	Canada	Human	—	MAI-HF	22/2725
469	"	New Jersey	Abstract	Wrapped cylinder	"	24/1949
470	"	Oklahoma	Abstract	Animal?	"	2/868
471	"	Canada	—	Perforated	"	2/9372
472	"	Connecticut?	—	—	AMNH	50.1/1624
473	"	Oklahoma	—	—	MAI-HF	16/9507
474	"	"	—	—	"	16/9460
475	"	Ontario, Can.	—	—	NYSM	37113
476	"	Moraviantown, Can.	—	—	MAI-HF	2/8472
477	"	—	—	—	AMNH	
478	Scaghticoke	New York	—	—	RMSC	AE 7296
479	Pequot	Connecticut	—	—	MAI-HF	10/7560

Text No.	Tribe	Location	Effigy	Notes	Coll.	Cat. No.
480	"	"	—	—	"	10/7558
481	"	"	—	—	"	10/7557
482	Nanticoke	Six Nations Res.	Swan/Goose	—	"	16/4195
483	Stockbridge	Wisconsin	—	—	"	24/2183
484	"	"	—	—	"	18/5917
485	Mohegan	Connecticut	Human	—	"	22/4306
486	"	"	—	—	"	3/5641B
487	"	"	—	—	"	3/5644A
488	"	"	—	—	"	3/5641A
489	"	"	—	—	"	3/5644B
490	"	"	—	—	"	12/9967
491	"	"	—	—	"	1/1025
492	"	"	—	—	"	10/8147
493	"	"	—	—	"	3/5641C
494	Mahican	New York	—	—	"	14/2099
495	"	"	—	—	"	14/2098
496	"	"	—	—	"	19/8286
497	Abenaki	Quebec	Turtle	Bas Relief	"	13/2804
498	"	"	—	Asymmetrical	"	11/6977
499	"	"	—	Perforated	"	11/6978
500	"	"	—	—	"	3/2435
501	Abenaki	Quebec	—	—	MAI-HF	11/6980
502	"	"	—	—	"	15/4312
503	"	"	—	—	"	18/488
504	"	"	—	—	MAI-HF	11/6979
505	"	"	—	—	"	12/5511
506	"	"	—	Cross Hatch Incising	"	14/3103
507	Cree	"	Beaver	Tetes de Boule	"	
508	"	"	Ball	—	BM	1494
509	"	"	—	—	"	1921.10.4.200
510	Naskapi	Canada	—	—	FMNH	176896
511	"	"	—	—	"	176899
512	"	"	—	—	"	176553
513	"	"	—	—	"	176900
514	"	"	—	—	"	176902
515	"	"	—	—	"	176901
516	"	"	—	—	PM-HU	87237A
517	"	"	—	—	FMNH	177324
518	Mistassini	Wisconsin	—	—	"	155867
519	"	"	—	—	"	155865
520	Potawatomi	Kansas	Bird Head	—	MAI-HF	2/7577C
521	"	—	Bear	—	Ritzenthaler	1976:40
522	"	Kansas	Bear	—	MAI-HF	2/7577B
523	"	Michigan	Animal Head	—	PM-HU	64729
524	"	Wisconsin	Abstract	—	MPM	23349
525	"	Michigan	"	—	MAI-HF	15/2838
526	"	Wisconsin	"	—	"	21/3395
527	"	"	"	—	MPM	23347
528	"	Kansas	"	—	MAI-HF	2/7577A
529	"	Wisconsin	—	—	FMNH	155693
530	"	"	—	—	MPM	55494
531	"	"	—	—	MAI-HF	20/5173
532	"	"	—	—	MPM	23346
533	"	"	—	—	MAI-HF	19/582
534	Potawatomi	Wisconsin	—	—	MPM	23348
535	"	"	—	"for medicine"	FMNH	155815
536	"	"	—	—	MPM	23337
537	"	"	—	—	"	23344
538	"	"	—	—	"	23345
539	"	"	—	—	FMNH	155813
540	"	"	—	—	"	155814
541	"	"	—	"for medicine"	"	155819

Text No.	Tribe	Location	Effigy	Notes	Coll.	Cat. No.
542	"	"	—	—	"	155851
543	"	Oklahoma	—	—	MAI-HF	1/9985
544	"	Kansas	—	—	"	2/7577E
545	"	"	—	Perforated	"	2/7577D
546	"	"	—	—	PM-HU	80352
547	"	"	—	—	MAI-HF	2/7577F
548	"	Michigan	—	—	PM-HU	64728
549	"	"	—	—	"	64731
550	"	"	—	—	"	64730
551	Winnebago	Wisconsin	Bird Head	Pop-Eyed, C-Shaped	FMNH	69294
552	"	"	"	"	AMNH	50/7648
553	"	"	"	C-Shaped	MPM	3443
554	"	"	"	"	FMNH	69292
555	"	"	"	"	MV-B	B 5759
556	"	"	"	"	MPM	56231
557	"	"	"	"	"	14650
558	"	"	"	"	AMNH	50.1/1410
559	"	"	"	"	MPM	3285
560	"	"	"	—	FMNH	69295
561	"	"	"	—	"	14890
562	"	—	"	—	PM-HU	62990
563	"	Wisconsin	"	C-Shaped	AMNH	50/7525
564	"	"	Pig Head	—	FMNH	69291
565	"	"	Eared Animal	"love medicine"	"	155962
566	"	Nebraska	Abstract	—	MAI-HF	16/9433
567	Winnebago	Wisconsin	Abstract	—	FMNH	14894
568	"	"	"	—	"	14892
569	"	—	—	—	PM-HU	73011
570	"	—	—	C-Shaped	AMNH	
571	"	—	—	—	"	50.1/896
572	"	Wisconsin	—	—	"	50/7511
573	"	"	—	—	"	50/7836
574	"	"	—	—	"	50/7512
575	"	"	—	—	FMNH	69289
576	"	"	—	—	"	14917
577	"	"	—	—	MPM	3282
578	"	"	—	—	"	3286
579	"	"	—	—	MV-B	B 5760
580	"	Indiana	—	—	PM-HU	65778
581	"	Wisconsin	—	—	FMNH	14889
582	"	"	—	—	AMNH	50/7644 (5?)
583	"	"	—	—	"	50/7698
584	"	"	—	—	MV-B	B 5758
585	"	Nebraska	—	—	FMNH	155641
586	"	"	—	—	AMNH	50.2/4475
587	Sauk & Fox	Oklahoma	Human	—	MAI-HF	2/6557A
588	"	"	Human	"used in Midewiwin"	"	2/5717
589	"	"	Finger	—	MPM	30283
590	"	"	"Eagle"	—	FMNH	34819
591	"	Iowa, Tama	"	—	"	34820
592	"	"	Bird Head	—	"	34827
593	"	Oklahoma	"	—	MAI-HF	2/5553B
594	"	"	Bear	Painted, Perforated	NMM	III-I-236
595	"	"	Pocupine	—	MPM	30282
596	"	"	Bear/Wolf	—	MAI-HF	2/7578
597	"	—	Horse	—	AMNH	
598	"	—	"	—	"	
599	"	Iowa, Tama	"	—	FMNH	34828
600	"	Oklahoma	Buffalo	—	MAI-HF	2/5722

Text No.	Tribe	Location	Effigy	Notes	Coll.	Cat. No.
601	Sauk & Fox	Oklahoma	Animal	—	MAI-HF	2/5553C
602	"	"	Turtle	"War Bundle Feast"	"	2/6555
603	"	—	—	—	FMNH	207744
604	"	Oklahoma	Abstract	—	MAI-HF	2/5553A
605	"	—	"	—	"	2/6557B
606	"	—	"	—	PM-HU	84031
607	"	Oklahoma	"	—	MAI-HF	2/5155A
608	"	"	—	—	MPM	30289
609	"	"	—	—	"	30288
610	"	"	—	—	"	30287
611	"	"	—	—	"	30286
612	"	"	—	—	"	30285
613	"	"	—	—	"	30284
614	"	"	—	—	"	30281
615	"	"	—	—	"	30290
616	"	"	—	—	MAI-HF	3/2698
617	"	"	—	Perforated	"	2/4837B
618	"	"	—	—	"	2/5155B
619	"	"	—	—	"	2/4837A
620	"	Iowa, Tama	—	—	MV-B	B 6398
621	"	"	—	—	"	B-6400
622	"	"	—	—	"	B 6399
623	"	"	—	—	FMNH	34824
624	"	"	—	—	"	34826
625	"	"	—	—	"	34821
626	"	"	—	—	"	34818
627	"	"	—	—	"	34823
628	"	"	—	—	"	34825
629	Menominee	Wisconsin	Bear/Wolf	—	AMNH	50/4836
630	"	"	Rabbit	—	"	50/9897
631	"	"	Animal	C-Shaped	MPM	4476
632	"	"	—	Perforated	"	4474
633	"	"	—	—	"	4473
634	Menominee	Wisconsin	—	—	"	4472
635	"	"	—	—	"	4471
636	"	"	—	—	"	4470
637	"	"	—	Perforated	"	4469
638	"	"	—	—	"	4468
639	"	"	—	—	FMNH	155950
640	"	"	—	—	"	155948
641	"	"	—	Perforated	"	155947
642	"	"	—	—	MPM	4477
643	"	"	—	Perforated	"	4475
644	"	"	—	—	MV-B	B 6432
645	"	"	—	—	MPM	4479
646	"	"	—	—	MV-B	B-6433
647	"	"	—	Perforated	MPM	28145
648	"	"	—	—	"	4480
649	"	"	—	—	"	59220
650	"	"	—	Top Broken	"	4478
651	Wyandot	Oklahoma	—	—	NMM	III-I-429
652	Mascouten	Kansas	Beaver	—	MPM	31328
653	"	"	Otter	—	"	31330
654	"	"	Ball Head Club	—	"	31324
655	"	"	—	—	"	31322
656	"	"	—	—	"	31323
657	"	"	—	—	"	31321
658	"	"	—	Perforated	"	31329
659	"	"	—	—	"	31333
660	Ojibway	—	Clenched Fist	—	PM-S	
661	"	Ont., Canada	"Crane"	—	PM-HU	10/12738
662	"	—	Bird	—	PM-S	
663	"	Ont., Canada	Bird	—	PM-HU	65071B

Text No.	Tribe	Location	Effigy	Notes	Coll.	Cat. No.
664	"	Walpole Is., Can.	Bird Head	—	"	65071A
665	"	Wisconsin	—	Perforated	MPM	4996
666	"	Walpole Is., Can.	—	—	PM-HU	64689
667	Ojibway	Walpole Is., Can.	—	—	"	65071E
668	"	"	—	—	"	64688
669	"	"	—	—	"	64687
670	"	"	—	—	"	65071D
671	"	"	—	—	"	65073B
672	"	"	—	—	"	65073A
673	"	"	—	—	"	65071F
674	"	Parry Is., Can.	—	—	RMSC	AE 1542B
675	"	"	—	—	"	AE 1542A
676	"	"	—	—	"	AE 1544
677	"	Lake Superior	—	—	PM-HU	10215
678	"	Ont., Canada	—	—	"	64750
679	"	Walpole Is., Can.	—	—	"	64690
680	"	Ont., Can.	—	Fancy Handle	"	64748
681	"	"	—	"	"	64753
682	"	"	—	"	"	64752
683	"	"	—	—	"	64754
684	"	"	—	—	"	65020
685	"	"	—	Leaves incised, perf.	"	64749
686	"	Walpole Is., Can.	—	—	"	65071C
687	"	Ont., Canada	—	Keyhole Perforation	"	64752B
688	"	Ont., Canada	—	Incised Perf.	"	64751
689	Chippewa	—	Long-Leg Bird	—	Ritzenthaler	1976:40
690	"	—	Bird	—	Maurer	1977:127
691	"	—	Bird Head	—	Ritzenthaler	1976:40
692	"	—	Ball Head Club	—	Densmore	1929:pl.17
693	"	Wisconsin	—	—	MPM	5008
694	"	"	—	—	"	52708
695	"	Canada	—	—	PRM	1939.6.B.15
696	Ottawa	—	Human	—	Ritzenthaler	1976:40
697	"	Lake Huron	Bird	—	BM	ST 792
698	"	—	Animal Head	—	MV-B	B 12857
699	"	Ont., Canada	—	—	FMNH	15379
700	"	Ont., Canada	—	—	RMSC	AE 1559

REFERENCES CITED

Abrams, George
    1965   The Cornplanter Cemetery. *Pennsylvania Archaeologist* 35 (2) August.

Adair, James
    1775   *The History of the American Indians, Particularly Those Nations Adjoining the Mississippi, East and West Florida, Georgia, South and North Carolina, and Virginia.* Printed for Edward and Charles Dilly, London.

Anonymous
    1600   The first voyage made to the coasts of America with two Barks, wherein were Captaines M. Philip Amadar and M. Arthur Barlove, who discovered part of the country now called Virginia. Anno 1584. In, *The Principall Navigations, Voiages, and Discoveries of the English Nation,* by Richard Hakluyt, Vol. 3. (G)eorge Bishop, Rolfe Newberie & Robert Barker, London. Cited in Peter P. Pratt, *Archaeology of the Oneida Iroquois, Vol. 1. Occasional Publications in Northeastern Anthropology,* No. 1. Man in the Northeast, George's Mills, New Hampshire.

Beauchamp, William M.
    1895   Onondaga Notes. *Journal of American Folklore.* 8(30):209-216.
    1901   The Good Hunter & the Iroquois Medicine. *Journal of American Folklore* 14:153-159.
    1905   Aboriginal Use of Wood in New York. *New York State Museum Bulletin* 89. Albany.
    1907   Civil, Religious & Mourning Councils and Ceremonies of Adoption of the New York Indians. *New York State Museum Bulletin* 113. Albany.
    1916   *Moravian Journals Relating to Central New York. 1745-66.* Dehler Press, Syracuse, New York.
    1922   *Iroquois Folk Lore, Gathered from the Six Nations of New York.* Port Washington, New York.

Beverley, Robert
    1722   *The History of Virginia.* (second ed.). Printed for B. & S. Tooke, London.

Biggar, H. P.
    1929   *The Works of Samuel de Champlain.* Vol. III The Champlain Society, Toronto.

Bossert, Helmuth T.
    1977   *Peasant Art of Europe and Asia.* Hastings House, New York.

Brasser, T. J.
    1977   North American Indian Art for T M. In, *The Religious Character of Native American Humanities.* Paper read at an interdisciplinary conference, April 14-15, 1977. Department of Humanities and Religious Studies, Arizona State University, Tempe, Arizona, pp. 12-43.
    1978   Mahican. In *Handbook of North American Indians, Northeast. Vol. 15.* Volume edited by Bruce G. Trigger. Smithsonian Institution, Washington, D. C.

Callender, Charles
    1978   Sauk. In, *Handbook of North American Indians, Northeast, Vol. 15.* Volume edited by Bruce G. Trigger. Smithsonian Institution, Washington, D. C.

Cameron, Donald
    n.d.   Unpublished Field Notes. On file at the Rochester Museum & Science Center, Rochester, New York.

Campbell, Patrick
    1978   A Journey through the Genesee Country, Finger Lakes Region and Mohawk Valley. From, *Patrick Campbell's Travels in the Interior Inhabited Parts of North America in the Years 1791 and 1792.* The Friends of the University of Rochester Libraries, Rochester, New York.

Chafe, Wallace L.
  1961 Comment on Anthony F. C. Wallace's 'Cultural Composition of the Handsome Lake Religion.' In, *Symposium on Cherokee and Iroquois Culture,* edited by William N. Fenton. *Bureau of American Ethnology Bulletin* 180:pp. 155-157. Smithsonian Institution, Washington, D. C.

Clark, Hugh
  1892 *An Introduction to Heraldry.* George Bell & Sons, London.

Coleman, Emma Lewis
  1925 *New England Captives Carried to Canada between 1677-1760, during the French & Indian Wars.* The Southern Press, Portland, Maine.

Curtin, Jeremiah
  1923 *Seneca Indian Myths.* New York.

Curtin, Jeremiah and J.N.B. Hewitt
  1918 Seneca Fiction, Legends, and Myths. Part I. *Bureau of American Ethnology Annual Report 32.* Smithsonian Institution, Washington, D. C.

Cushing, Frank Hamilton
  1896 Exploration of Ancient Key Dwellers' Remains on the Gulf Coast of Florida. *Proceedings of the American Philosophical Society,* 25 (153). Quoted in *Sun Circles & Human Hands* by Emma Lila Fundaburk, Ed. 1957. Luverne, Alabama.

Deardorff, Merle & G. S. Snyderman
  1956 A 19th Century Journal of a Visit to the Indians of New York (Written by Quaker Missionaries John Philips, Halliday Jackson & Isaac Bonsal. *American Philosophical Society, Proceedings* 100(6).

Densmore, Frances
  1929 Chippewa Customs. *Bureau of American Ethnology Bulletin 86.* Smithsonian Institution, Washington, D.C.
  1948 A Collection of Specimens from the Teton Sioux. *Indian Notes & Monographs* 11(3). Museum of the American Indian, Heye Foundation.

Denton, Daniel
  1902 *A Brief Description of New York formerly called New Netherlands.* Reprinted from the original edition of 1670. The Burrows Brothers Company, Cleveland.

Dockstader, Frederick J.
  1961 *Indian Art in North America. Arts & Crafts.* New York Graphic Society, Greenwich, Connecticut.
  1973 *Indian Art of the Americas.* Museum of the American Indian, Heye Foundation, New York.

Dodge, Ernest S.
  1951 Some Thoughts on the Historic Art of the Indians of Northeastern North America. *Massachusetts Archeological Society Bulletin* 13(1).

Dorsey, James O.
  1894 A Study of Siouan Cults. *Bureau of American Ethnology Annual Report 11* Smithsonian Institution, Washington D.C.

Eliade, Mircea
  1964 *Shamanism, Archaic Techniques of Ecstasy.* Translated from the French by Willard R. Trask. Bollingen Series LXXVI Pantheon Books.

Fenton, William N.
  1936 An Outline of Seneca Ceremonies at Coldspring Longhouse. *Yale University Publications in Anthropology* No. 9.
  1940 Problems Arising from the Historic Northeastern Position of the Iroquois. In, *Essays in Historical Anthropology of North America. Smithsonian Miscellaneous Collections* 100, Washington, D.C.
  1941 Masked Medicine Societies of the Iroquois. *Smithsonian Institution Annual Report for* 1940, Washington, D.C.
  1942 Songs from the Iroquois Longhouse: Program Notes for an Album of American Indian Music from the Eastern Woodlands. Publication 3691. Smithsonian Institution, Washington, D. C.112

1950 The Roll Call of the Iroquois Chiefs. A Study of a Mnemonic Cane from the Six Nations Reserve. *Smithsonian Miscellaneous Collections.* Vol. 3, No. 15. Washington, D. C.

1951 Locality as a Basic Factor in the Development of Iroquois Social Structure. *Bureau of American Ethnology Bulletin* 149(3):35-54. Washington, D. C.

1961 Iroquoian Culture History: A General Evaluation. *Bureau of American Ethnology Bulletin* 180. Washington, D. C.

Fenton, William N. & Elizabeth L. Moore, translators & editors.
1974-7 *Customs of the American Indians Compared with the Customs of Primitive Times by Father Joseph Francois Lafitau* (1724). Volumes 1-2. Publications of The Champlain Society, Volumes 48-9, Toronto, Canada.

Flayderman, E. Norman
1972 *Scrimshaw and Scrimshanders, Whales & Whalemen.* N. Flayderman & Co., Inc., New Milford, Conn.

Friedmann, Herbert
1946 *The Symbolic Goldfinch.* Pantheon Books.

Gehring, Charles
Personal Communication, New York State Library June, 1979.

Gookin, Daniel
1970 *Historical Collections of the Indians in New England* (1792), edited by Jeffrey H. Fiske. Towtaid.

Grant, Francis J. (editor)
1924 *The Manual of Heraldry.* Edinburgh.

Grant, W. L. (editor)
1907 *Voyages of Samuel de Champlain 1604-1618.* Charles Scribner's Sons, New York.

Gunn, Sarah
1782 *Captivity of Sarah Whitmore*

Hallowell, A. Irving
1926 Bear Ceremonialism in the Northern Hemisphere. *American Anthropologist* n.s. Vol. 28, No. 1.

Hamell, George R.
1979a The Genesee Valley Seneca Iroquois, 1750-97; Contemporary Sources. In *A Genesee Harvest. A Scene in Time. 1779.* Genesee Valley Council on the Arts. Watkins Glen, N.Y. Walnut Grove Design & Production Associates.

1979b *Of Hockers, Diamonds & Hourglasses, Some Interpretations of Seneca Archaeological Art.* Paper presented at the Iroquois Conference, Albany, N.Y., October 13-15, 1979.

1980a Gannagaro State Historic Site: A Current Perspective. *Occasional Publications in Northeastern Anthropology. Franklin Pierce College. Rindge, NH.*

1980b Sun Serpents, Tawiskaron & Quartz Crystals. Paper read at Annual Conference on Iroquois Research. Rensselaerville, N.Y. October 10-12, 1980.

Hanson, Charles, Jr.
1975 The Crooked Knife. *Museum of the Fur Trade Quarterly.* 11(2) Summer.

Harrington, Mark R.
1908 Vestiges of Material Culture among the Canadian Delawares. *American Anthropologist.* n.s. 10(3): 408-418.

Hayes, Charles F. III
1966 Pits of the Archaic Stage Salvaged from the Farrell Farm. *Museum Service. Bulletin of the Rochester Museum of Arts & Sciences.* 39(9-10): 167-175.

Hayes, Charles F. III & Lilita Bergs
1969 A Progress Report on an Archaic Site on the Farrell Farm, the Cole Gravel Pit, 1966-1968. *New York State Archaeological Association Bulletin* 47: 1-12.

Hendry, Jean
1964 Iroquois Mask and Maskmaking at Onondaga. *Bureau of American Ethnology Bulletin* 191. Washington, D.C.

Hennepin, Louis
1903 *A New Discovery of a Vast Country in America.* (1698). Edited by Reuben G. Thwaites, 2 Vols. A. C. McClurg, Chicago.

Hewitt, J. N. B.
   1903   Iroquoian Cosmology. First Part, *21st Annual Report of the Bureau of American Ethnology to the Secretary of the Smithsonian Institution 1899-1900.* Washington, D.C. 127-339.

   1928   Iroquoian Cosmology. Second Part, *43rd Annual Report of the Bureau of American Ethnology to the Secretary of the Smithsonian Institution 1925-26.* Washington, D.C., 449-819.

Hoffman, William J.
   1891   The Midewiwin or 'Grand Medicine Society,' *7th Annual Report of the Bureau of Ethnology to the Secretary of the Smithsonian Institution 1885-1886.* Washington, D.C. 143-300.

Holmes, W. H.
   1883   Art in Shell of the Ancient Americans. *2nd Annual Report of the Bureau of American Ethnology to the Secretary of the Smithsonian Institution.* Washington, D.C.

Jesuit Relations (JR)
   See Thwaites, Reuben Gold, ed.

Jackson, Halliday
   1830   *Civilization of the Indian Natives.* Philadelphia.

Jameson J. Franklin (editor)
   1909   *Narratives of New Netherlands, 1609-1664.* Charles Scribner's Sons, New York. Reprinted. Barnes & Noble, New York 1959.

deJonge, Eric, ed.
   1973   *Country Things.* The Pyne Press, Princeton, New Jersey.

King, Jonathan C. H.
   1977   *Smoking Pipes of the North American Indian.* British Museum Publications, Ltd., London.

Kinietz, W. Vernon
   1940   The Indians of the Western Great Lakes, 1615-1760. *Occasional Contributions* 10, University of Michigan Museum of Anthropology, Ann Arbor.

Klees, Fredric
   1950   *The Pennsylvania Dutch.* The MacMillan Company, New York.

de Lahontan, Baron Louis Armand
   1905   *New Voyages to North-America. Vol. II.* (1703) A. C. McClurg & Co., Chicago.

Lawson, John
   1709   *A New Voyage to Carolina; Containing the Exact Description and Natural History of That Country, Together with the Present State Thereof and a Journal of a Thousand Miles Traveled Through Several Nations of Indians, Giving a Particular Account of Their Customs, Manners, etc.* London: (no publisher).

Levi-Strauss, Claude
   1978   *The Origin of Table Manners.* Translated from the French by John & Doreen Weightman. Harper & Row, New York.

Loskiel, George H.
   1794   *History of the Mission of the United Brethren Among Indians in North America.* Translated by Christian Ignatius La Trobe. 3 pts. Printed for the Brethren's Society for the Furtherance of the Gospel, London.

MacDonald, George F.
   1977   The Problems & Promise of Wet Site Archaeology. CCI, *The Journal of the Canadian Conservation Institute. Vol. 2.*

MacNeish, Richard S.
   1952   Iroquois Pottery Types, a Technique for the Study of Iroquois Prehistory. *Bulletin.* 124. *National Museum of Canada,* Ottawa, Canada.

Mallery, Garrick
   1893   Picture Writing of the American Indians. *10th Annual Report of the Bureau of Ethnology to the Secretary of the Smithsonian Institution 1888-1889.* Washington, D.C. 1-807.

Mathews, Zena Pearlstone
   1978   *The Relation of Seneca False Face Masks to Seneca and Ontario Archeology.* Garland Publishing Inc., New York.

Maurer, Evan N.
    1977   *The Native American Heritage. A Survey of North American Indian Art.* The Art Institute of Chicago. University of Nebraska Press, Lincoln.

Megapolensis, Johannes, Jr.
    1909   A Short Account of the Mohawk Indians . . . 1644. In, *Narratives of New Netherland, 1609-1664,* edited by J. Franklin Jameson. Charles Scribner's Sons, New York.

Miniter, Edith
    1973   When Treen Ware Was 'The' Ware, in *Country Things,* edited by Eric deJonge. The Pyne Press. Princeton, New Jersey.

Morgan, Lewis Henry
    1852   Report on the Fabrics, Inventions, Implements, and Utensils of the Iroquois. *New York State Cabinet of Antiquities' Annual Report* 5:66-117. Albany, New York.

    1962   *League of the Ho-De-No-Sau-Nee, Iroquois.* Corinth Books, New York (Rochester, N.Y.: Sage & Brother, 1851).

Parker, Arthur C.
    1908   Myths & Legends of the New York Iroquois. *New York State Museum Bulletin* 125. Albany.

    1909   Secret Medicine Societies of the Seneca. *American Anthropologist* n.s. 11(2):161-185. (Reprinted in Parker 1913:113-130).

    1910   Iroquois Uses of Maize and Other Food Plants. *New York State Museum Bulletin* 144. Albany.

    1913   The Code of Handsome Lake, the Seneca Prophet. *New York State Museum Bulletin* 163. Albany.

    1916   Constitution of the Five Nations. *New York State Museum Bulletin* 184. Albany.

    1923   Seneca Myths and Folk-tales. *Buffalo Historical Society. Proceedings* 27:465p.

Pastorius, Francis D.
    1912   *Narratives of Early Pennsylvania, West New Jersey, and Delaware, 1630-1707.* Edited by Albert C. Myers. Charles Scribner's Sons, New York.1

Pratt, Peter P.
    Archaeology of the Oneida Iroquois, Vol. 1. Occasional Publications in Northeastern Anthropology, No. 1. Man in the Northeast, George's Mills, NH.

Quimby, George Irving
    1966   *Indian Culture and European Trade Goods.* University of Wisconsin Press, Madison.

Rainwater, Dorothy T. and Donna H. Felger
    1976   *A Collector's Guide to Spoons Around the World.* Everybody's Press, Inc., Hanover, Pennsylvania.

Randle, Martha Champion
    1953   The Waugh Collection of Iroquois Folktales. *American Philosophical Society Proceedings* 97:611-633.

Ritchie, William A.
    1932   The Lamoka Lake Site: The Type Station of the Archaic Algonkin Period in New York. *Researches and Transactions of the New York State Archeological Association* 7(4). Rochester, New York.

    1936   A Prehistoric Fortified Village Site at Canandaigua, Ontario County, New York. *Research Records of the Rochester Museum of Arts and Science No.* 3. Rochester.

    1945   An Early Site in Cayuga County, New York. Type Component of the Frontenac Focus, Archaic Pattern. *Research Records of the Rochester Museum of Arts & Science. No.* 7. Rochester.

    1947   Archaeological Evidence for Ceremonialism in the Owasco Culture. *Researches & Transactions of the New York State Archeological Association* 11(2) Rochester.

    1950   Another Probable Case of Prehistoric Bear Ceremonialism in New York. *American Antiquity* 15(3). January.

    1965   *The Archaeology of New York State.* Natural History Press, Garden City, New York.

Ritzenthaler, Robert E.
    1976 Woodland Sculpture. *American Indian Art.* Autumn.
    1978 Southwestern Chippewa. In, *Handbook of the North American Indians, Northeast. Vol. 15.* Volume edited by Bruce G. Trigger. Smithsonian Institution, Washington, D.C.

Scribner, Charles (editor)
    1917 *Early Narratives of the Northwest.* Charles Scribner's Sons, New York.

Seaver, James E.
    1967 *A Narrative of the Life of Mrs. Mary Jemison.* Facsimile Copy. Charles C. Kelsey & Allegany Press.

Shimony, Annemarie A.
    1961 Conservatism Among the Iroquois at the Six Nations Reserve. *Yale University Publications in Anthropology* 65. New Haven, Connecticut.

Skinner, Alanson
    1920a Medicine Ceremony of the Menomini, Iowa, and Wahpeton Dakota, with Notes on the Ceremony Among the Ponca, Bungi Ojibwa, and Potawatomi. *Indian Notes & Monographs, vol. IV.* Museum of the American Indian, Heye Foundation. New York.
    1920b Two Antler Spoons from Ontario. *Indian Notes and Monographs.* Museum of the American Indian, Heye Foundation. New York.

Smith, James
    1799 *An Account of the Remarkable Occurrences in the Life and Travels of Col. James Smith, During His Captivity with the Indians in the Years 1755, '56, '57, '58, and '59...*John Bradford, Lexington, Virginia.

Speck, Frank G.
    1909 Ethnology of the Yuchi Indians. *Anthropological Publications.* University Museum. University of Pennsylvania. Vol. 1. No. 1. Philadelphia. Cited in Swanton, John R. The Indians of the Southeastern United States. *Bureau of American Ethnology Bulletin* 137:1946.
    1915 *Decorative Art of the Indian Tribes of Connecticut.* Government Printing Bureau, Ottawa.
    1937 Oklahoma Delaware Ceremonies, Feasts & Dances. *Memoirs of the American Philosophical Society Vol. VII.* Philadelphia.
    1945 The Iroquois: A Study in Cultural Evolution. *Cranbrook Institute of Science Bulletin* 23. Bloomfield Hills, Michigan.

Sturtevant, William C.
    1978 Oklahoma Seneca-Cayuga. In *Handbook of North American Indians, Northeast, Vol. 15.* Edited by Bruce G. Trigger. Washington, D.C.: Smithsonian Institution, Washington, D.C.
    1979 Louis Philippe on Cherokee Architecture and Clothing in 1797. *Journal of Cherokee Studies.* Vol. III, No. 3, Fall.

Swanton, John
    1946 The Indians of the Southeastern United States. *Bureau of American Ethnology Bulletin* 137. Washington, D.C.

Thompson, Stith
    1966 *Tales of the North American Indians.* Indiana University Press, Bloomington.

Thwaites, Reuben Gold (editor)
    1896-1901 *The Jesuit Relations and Allied Documents.* Volumes 1-73. Burrows Brothers Company, Cleveland.

Trigger, Bruce G., ed.
    1978 *Handbook of North American Indians, Northeast Vol. 15.* Smithsonian Institution, Washington, D.C.

Ubbelohde-Doering, Heinrich
    1954 *The Art of Ancient Peru.* New York: Frederick A. Praeger.

Vastokas, Joan M. & Romas K. Vastokas
    1973 *Sacred Art of the Algonkians.* Mansard Press, Peterborough, Canada.

Ventur, Pierre
    1980 A Comparative Perspective on Native American Mortuary Games of the Eastern Woodlands. *Man in the Northeast Vol. 20.*

Wallace, Anthony F. C.
   1972   *The Death & Rebirth of the Seneca.* Vintage Books Edition. Random House, Inc., New York. Originally published by Alfred A. Knopf, Inc. 1970.

Wallace, Paul A. W.
   1951   They Knew the Indian; The Men Who Wrote the Moravian Records. *Proceedings of the American Philosophical Society* 95 (3): 290-295.
   1958   *Thirty Thousand Miles with John Heckewelder.* University of Pittsburgh Press.

Waring, A. J., Jr. and Preston Holder
   1945   A Prehistoric Ceremonial Complex in the Southeastern United States. *American Anthropologist* 47(1).

Waugh, Frederick W.
   1916   Iroquois Foods and Food Preparation. *Anthropological Series* 12. *Memoirs of the Canadian Geological Survey* 86. Ottawa.

Whitney, Theodore
   1974   Wooden Artifacts of New York State Indians *Chenango Chapter New York State Archeological Association* 15(4).

Willey, Gordon R.
   1966   *An Introduction to American Archaeology, Vol. 1: North & Middle America.* Prentice-Hall, Inc., Englewood Cliffs, New Jersey.

Willoughby, Charles C.
   1935   *Antiquities of the New England Indians.* Published by the Peabody Museum of American Archaeology and Ethnology. Harvard University, Cambridge, Massachusetts.

Wilson, James G.
   1895   Arent Van Curler and his Journal of 1634-35. *Annual Report of the American Historical Association.*:81-101. Washington, D.C.

Wood, Alice S.
   1974   A Catalogue of Jesuit and Ornamental Rings from Western New York: Collections of Charles F. Wray and the Rochester Museum and Science Center. *Historical Archaeology.* Vol. VIII pp. 83-104.

Wray, Charles F.
   1964   The Bird in Seneca Archeology. *Rochester Academy of Science Proceedings Vol.* 11.
   1973   *Manual for Seneca Iroquois Archeology.* Cultures Primitive, Inc., Honeoye Falls, New York.
   n.d.   Unpublished Field Notes. On file at the Rochester Museum & Science Center, Rochester.

Wray, Charles F. and Harry L. Schoff
   1953   A Preliminary Report on the Seneca Sequence in Western New York 1550-1687. *Pennsylvania Archeologist* 23(2).

Wrong, George M. (editor)
   1939   *The Long Journey to the Country of the Hurons by Father Gabriel Sagard.* The Champlain Society, Toronto.

Zeisberger, David
   1910   History of the Northern American Indians. Edited by Archer B. Hulbert & William N. Schwarze. *Ohio State Archaeological & Historical Quarterly* 19:1-189. Columbus.

CHART 1
17th CENTURY SENECA IROQUOIS ARCHAEOLOGICAL LADLES

	Cameron	Factory Hollow	Warren	Steele	Power House	Marsh	Dann	Markham	Rochester Junction	Boughton Hill	Beale	17th Century Total
Human			1	3		4	1		1			10
Fist/Hand/Finger				2		1	1			1		5
Reclining Figure						1	1			1		3
Leg & Foot												0
Human(s) in Scene												0
Human & Mask												0
Human & Bear												0
Human & Buffalo												0
Human & Turtle												0
Bear(s) or Wolf		1	1	1	2	3			1	3	1	13
Panther						1						1
Beaver							2					2
Otter(s)					1							1
Sheep												0
Horse				1								1
Buffalo												0
Cat												0
Pig												0
Rabbit												0
Squirrel												0
Porcupine												0
Skunk												0
Double Animal												0
Animal												0
Animal Head												0
Animal & Swan												0
Hawk or Eagle			1			1			1			3
Owl								1				1
Multiple Ducks												0
Swan or Goose												0
Heron/Snipe/Crane						1						1
Perching Bird												0
Bird Head												0
Turtle				2		1	1			3		7
Frog												0
Snake				1								1
Hourglass						1						1
Ball or Disc												0
Ball-Headed Club												0
Abstract			1		1	2	1	2		1		8
Non-Effigy	2	1			1	1	3					8
Top Broken		4		3	1	3	6		1	5	1	24
Total	2	7	3	14	7	19	17	1	4	14	2	90

CHART 2
18th CENTURY SENECA IROQUOIS ARCHAEOLOGICAL LADLES
+1 ETHNOLOGICAL LADLE

	Snyder-McClure	Huntoon	Townley-Read	Honeoye	Kendaia	Fall Brook	Big Tree	Canawaugus	Cattaraugus (Ethnol.)	Total 18th Century
Human		1								1
Fist/Hand/Finger										0
Reclining Figure	1									1
Leg & Foot										0
Human(s) in Scene										0
Human & Mask										0
Human & Bear				1						1
Human & Buffalo										0
Human & Turtle										0
Bear(s) or Wolf	1							1		2
Panther										0
Beaver						1				1
Otter(s)						1				1
Sheep										0
Horse										0
Buffalo										0
Cat										0
Pig										0
Rabbit										0
Squirrel										0
Porcupine										0
Skunk										0
Double Animal										0
Animal										0
Animal Head										0
Animal & Swan										0
Hawk or Eagle						1				1
Owl							1			1
Multiple Ducks										0
Swan or Goose					1					1
Heron/Snipe/Crane										0
Perching Bird									1	1
Bird Head										0
Turtle	1									1
Frog										0
Snake										0
Hourglass	1									1
Ball or Disc										0
Ball-Headed Club										0
Abstract	1	1								2
Non-Effigy										0
Top Broken	3	1	1	1		2				8
Total	8	3	1	2	1	3	2	2	1	23

CHART 3
19th CENTURY NEW YORK & OKLAHOMA SENECA ETHNOLOGICAL LADLES

	Seneca-Cattaraugus	Seneca-Tonawanda	Cattaraugus or Tonawanda	Seneca-Allegany	Seneca	Total N.Y. Seneca	Oklahoma Seneca	Oklahoma Iroquois
Human	3		1		1	5	1	
Fist/Hand/Finger						0		
Reclining Figure						0		
Leg & Foot					1	1		
Human(s) in Scene	1	2			2	5		
Human & Mask					1	1	1	
Human & Bear						0		
Human & Buffalo						0		
Human & Turtle						0		
Bear(s) or Wolf						0	4	1
Panther						0		
Beaver						0		
Otter(s)						0	1	
Sheep						0		
Horse						0		
Buffalo	1					1	1	
Cat		1				1		
Pig						0		
Rabbit						0		
Squirrel					1	1		
Porcupine						0		
Skunk						0		
Double Animal	1				1	2		
Animal						0		
Animal Head	3	2	1			6	2	
Animal & Swan						0		
Hawk or Eagle						0		
Owl	1		1			2		
Multiple Ducks						0		
Swan or Goose	2	1			3	6	3	
Heron/Snipe/Crane			1			1		
Perching Bird	30	1	11	6	26	74	2	
Bird Head	2					2	3	
Turtle						0		
Frog					1	1		
Snake						0		
Hourglass						0		
Ball or Disc			1		1	2		
Ball-Headed Club						0		
Abstract			1		1	2		
Non-Effigy	11	1	3	1	6	22	8	2
Top Broken						0		
Total	55	8	20	7	45	135	26	3

CHART 4
19th CENTURY IROQUOIS (EXCEPT SENECA) ETHNOLOGICAL LADLES

	Cayuga	Onondaga	Oneida	Mohawk	Iroquois New York	Iroquois Grand River	Iroquois Canada	Iroquois	Total
Human	2	4	2					3	11
Fist/Hand/Finger			1			1			2
Reclining Figure									0
Leg & Foot									0
Human(s) in Scene									0
Human & Mask									0
Human & Bear									0
Human & Buffalo									0
Human & Turtle									0
Bear(s) or Wolf		1	1			1		1	4
Panther									0
Beaver				1					1
Otter(s)									0
Sheep		1							1
Horse								1	1
Buffalo		1					1	1	3
Cat						1			1
Pig									0
Rabbit		1						1	2
Squirrel			1						1
Porcupine									0
Skunk									0
Double Animal							2		2
Animal									0
Animal Head				1				2	3
Animal & Swan		2							2
Hawk or Eagle									0
Owl		1						1	2
Multiple Ducks	3								3
Swan or Goose		2	1		1		3	4	11
Heron/Snipe/Crane			3						3
Perching Bird	2	5	1		9	7	1	7	32
Bird Head	1	2			1	2		1	7
Turtle	1	1							2
Frog									0
Snake	1					1			2
Hourglass									0
Ball or Disc		1						1	2
Ball-Headed Club									0
Abstract	1	3	1		2		1	1	9
Non-Effigy	7	1	5	5	6	6	7	6	43
Top Broken									0
Total	18	26	16	7	19	19	15	30	150

CHART 5
19th CENTURY ETHNOLOGICAL LADLES
ACCORDING TO LOCATION WHEN COLLECTED
MIXED TRIBAL GROUPINGS

	Cattaraugus Reservation	Six Nations Reserve	Oklahoma	Wisconsin
Human	5	1	3	
Fist/Hand/Finger		2		
Reclining Figure				
Leg & Foot				
Human(s) in Scene				
Human & Mask				
Human & Bear				
Human & Buffalo	1			
Human & Turtle			1	
Bear(s) or Wolf		2	4	
Panther				
Beaver				
Otter(s)			1	
Sheep				
Horse				
Buffalo	1		2	
Cat		1		
Pig				1
Rabbit				1
Squirrel				1
Porcupine				
Skunk				1
Double Animal	1			
Animal			1	
Animal Head	3		4	2
Animal & Swan				
Hawk or Eagle				
Owl	1			
Multiple Ducks		3		
Swan or Goose	1	1	3	1
Heron/Snipe/Crane				3
Perching Bird	30	10	2	
Bird Head	2	1	4	10
Turtle		1	1	
Frog				
Snake		2		
Hourglass				
Ball or Disc				
Ball-Headed Club				
Abstract		3	3	8
Non-Effigy	11	15	30	55
Top Broken				
Total	56	42	59	83

CHART 6
19th CENTURY ETHNOLOGICAL LADLES, GREAT LAKES TRIBES

	Potawatomi	Winnebago	Sauk & Fox	Menominee	Wyandot	Mascouten	Ojibway	Chippewa	Ottawa	Total
Human			2						1	3
Fist/Hand/Finger	1		1				1			3
Reclining Figure										0
Leg & Foot										0
Human(s) in Scene										0
Human & Mask										0
Human & Bear										0
Human & Buffalo										0
Human & Turtle										0
Bear(s) or Wolf	2		1	1						4
Panther										0
Beaver							1			1
Otter(s)							1			1
Sheep										0
Horse			3							3
Buffalo			1							1
Cat										0
Pig		1								1
Rabbit				1						1
Squirrel										0
Porcupine			1							1
Skunk		1								1
Double Animal										0
Animal			1							1
Animal Head	1	1	1						1	4
Animal & Swan										0
Hawk or Eagle										0
Owl										0
Multiple Ducks										0
Swan or Goose										0
Heron/Snipe/Crane							1	1		2
Perching Bird							2	1	1	4
Bird Head	1	12	4				1	1		19
Turtle			1							1
Frog										0
Snake										0
Hourglass										0
Ball or Disc										0
Ball-Headed Club						1		1		2
Abstract	5	3	4	1						13
Non-Effigy	22	17	21	18	1	5	24	3	2	113
Top Broken										0
Total	32	35	41	21	1	8	29	7	5	179

CHART 7
19th CENTURY ALGONQUIAN ETHNOLOGICAL LADLES

	EASTERN	Delaware	Pequot	Scaghticoke	Nanticoke	Stockbridge	Mahican	Mohegan	Total	NORTHERN	Mistassini	Naskapi	Abenaki	Cree	Total
Human		1						1	2						0
Fist/Hand/Finger									0						0
Reclining Figure									0						0
Leg & Foot									0						0
Human(s) in Scene									0						0
Human & Mask									0						0
Human & Bear									0						0
Human & Buffalo									0						0
Human & Turtle									0						0
Bear(s) or Wolf									0						0
Panther									0						0
Beaver									0					1	1
Otter(s)									0						0
Sheep									0						0
Horse									0						0
Buffalo									0						0
Cat									0						0
Pig									0						0
Rabbit									0						0
Squirrel									0						0
Porcupine									0						0
Skunk									0						0
Double Animal									0						0
Animal									0						0
Animal Head									0						0
Animal & Swan									0						0
Hawk or Eagle									0						0
Owl									0						0
Multiple Ducks									0						0
Swan or Goose					1				1						0
Heron/Snipe/Crane									0						0
Perching Bird									0						0
Bird Head									0						0
Turtle									0				1		1
Frog									0						0
Snake									0						0
Hourglass									0						0
Ball or Disc									0						0
Ball-Headed Club									0						0
Abstract		2							2					1	1
Non-Effigy		7	3	1		2	3	8	24		2	8	9	1	20
Top Broken															
Total		10	3	1	1	2	3	9	29		2	8	10	3	23